COMPUTING AND ANALYSING ENERGY MINIMISATION PROBLEMS IN LIQUID CRYSTALS

Implementation using Firedrake

COMPUTING AND ANALYSING ENERGY MINIMISATION PROBLEMS IN LIQUID CRYSTALS
Implementation using Firedrake

JINGMIN XIA

National University of Defense Technology, China

World Scientific

EW JERSEY · LONDON · SINGAPORE · BEIJING · SHANGHAI · HONG KONG · TAIPEI · CHENNAI · TOKYO

Published by

World Scientific Publishing Co. Pte. Ltd.

5 Toh Tuck Link, Singapore 596224

USA office: 27 Warren Street, Suite 401-402, Hackensack, NJ 07601

UK office: 57 Shelton Street, Covent Garden, London WC2H 9HE

Library of Congress Control Number: 2023004639

British Library Cataloguing-in-Publication Data
A catalogue record for this book is available from the British Library.

COMPUTING AND ANALYSING ENERGY MINIMISATION PROBLEMS IN LIQUID CRYSTALS
Implementation using Firedrake

ISBN 978-981-127-338-4 (hardcover)
ISBN 978-981-127-339-1 (ebook for institutions)
ISBN 978-981-127-340-7 (ebook for individuals)

For any available supplementary material, please visit
https://www.worldscientific.com/worldscibooks/10.1142/13335#t=suppl

Typeset by Stallion Press
Email: enquiries@stallionpress.com

This book is dedicated to

my beloved family

for their boundless love and company

Preface

Liquid crystals are widely used in display devices and their indispensable applications have driven more than a century of scientific investigations. They are of great interest in physics, for their striking defect structures, e.g., defect walls and focal conics in smectics; and in mathematics, for the questions arising in their modelling and analysis. Two successful mathematical theories are the Oseen–Frank (vector-based) and Landau–de Gennes (tensor-based) theories for nematics (of course, there are also other theories from different modelling viewpoints). In the former, the order parameter is simple but a nonlinear constraint must be enforced in the optimisation. The latter theory becomes more appealing in characterising complex defects, as it supports experimentally-observed defects that Oseen–Frank does not. However, when it comes to the phenomenological modelling of other phases of liquid crystals such as smectics, mathematical theories have not been extensively studied.

This book takes a step forward in understanding several modelling and implementation issues related to different phases of liquid crystals: cholesterics, ferronematics and smectics. Moreover, it introduces recent developments in computing and analysing the energy minimising problems arising from liquid crystals. The methods are presented using standard theories from partial differential equations and finite element analysis. Several implementation examples are given in detail with algorithm descriptions using the open-source libraries Firedrake and Defcon. Students and researchers with mathematics, physics or material sciences backgrounds may benefit from this book. Those readers who are less familiar with the topic or implementation code are provided with relevant overviews and provided coding examples in the book.

For concreteness, this book focuses on three different phases of liquid crystals, leading to three corresponding parts. In the first part, we propose an augmented Lagrangian-type preconditioner to construct efficient solvers for Oseen–Frank problems arising in cholesterics (and nematics). We analyse two advantages of the augmented Lagrangian formulation: (i) it helps in controlling the Schur complement matrix, enabling the development of block preconditioners; (ii) it improves the discrete enforcement of the unit-length constraint of the director. Since the augmentation makes the director block harder to solve, we develop a robust multigrid algorithm for the augmented block. The resulting preconditioner is an efficient and robust approach for solving director-based models of liquid crystals.

The second part is devoted to investigating defect structures (e.g., jumps of the director and magnetisation vector) in ferronematics, through numerical bifurcation analysis. Novel bifurcations of the ferronematic problem of interest are studied to give a more complete picture of solution landscapes as the parameter space varies. The reported numerical results validate the corresponding theoretical analysis in Dalby *et al.* (2022), and show us the potential of the Landau–de Gennes theory in characterising complicated defects.

Convinced by the successful application of the Landau–de Gennes model in ferronematics, we move to developing effective models of smectic-A liquid crystals in the third part of this thesis. We propose a new continuum model, solving for a real-valued smectic order parameter for the density variation and a tensor-valued nematic order parameter for the director orientation. This expands on an idea mentioned by Ball and Bedford (2015). The model is challenging to discretise due to the high regularity of the density variation; to address this, a continuous interior penalty discretisation is employed. Numerical analysis and experiments are performed to confirm the effectiveness of the proposed model and discretisation. The model numerically captures important defect structures in focal conic domains and oily streaks (two typical defects in smectics).

There are many excellent books on liquid crystals and this book is certainly only a glimpse of recent progress from the viewpoint of combining mathematics and numerics for liquid crystals. Although the content is only limited to certain topics for concreteness, this book is a first attempt of the author to give rather recent efforts on understanding liquid crystal theories and numerics. Moreover, the author is grateful for all those pioneers and collaborators working on liquid crystals, thus making this book possible.

Contents

Preface   vii

List of Figures   xiii

List of Tables   xix

1. Introduction   1

    1.1   Background   1
    1.2   Some Common Notation   4
    1.3   Common Solver Details   5

Part 1: Nematic and Cholesteric Liquid Crystals   **7**

2. A Mathematical Model of Nematics and Cholesterics   9

    2.1   The Oseen–Frank Model   9
    2.2   Lagrange Multiplier and Newton Linearisation   14
    2.3   Augmented Lagrangian Form   20
           2.3.1   Penalising the constraint   20
           2.3.2   Approximation to the Schur complement   23
           2.3.3   Improvement of the constraint   29
    2.4   Summary   32

3. A Robust Multigrid Algorithm for the Augmented
    Director Block   33

    3.1 Overview on Multigrid Methods   34
         3.1.1 Solving linear equations   34
         3.1.2 Solving nonlinear equations   40
    3.2 Relaxation   43
         3.2.1 Robustness analysis of the approximate kernel . .   46
    3.3 Prolongation   51
    3.4 Summary   52

4. Numerical Experiments for Nematics and Cholesterics   53

    4.1 Algorithm Details   53
         4.1.1 Newton–Kantorovich method   54
    4.2 Numerical Results   55
         4.2.1 Periodic boundary condition in a square slab . . .   55
         4.2.2 Equal-constant nematic case in an ellipse   60
    4.3 Summary   62

Part 2: Ferronematic Liquid Crystals   65

5. A Mathematical Model of Ferronematics   67

    5.1 The Landau–de Gennes Model   68
    5.2 Full Model of Ferronematics   69
    5.3 Reduced Model: Order Reconstruction   76
    5.4 Summary   80

6. Computing Multiple Solutions Using Deflation   81

    6.1 Implementation Examples   81
         6.1.1 Two-dimensional examples   82
         6.1.2 Three-dimensional examples   84
    6.2 Deflation   88
    6.3 A Hyperelastic Example   90
         6.3.1 Saint Venant–Kirchhoff hyperelastic model   90
         6.3.2 Enforcement of the essential boundary
                condition   93
         6.3.3 Bifurcation analysis of buckling behaviours   95
    6.4 Summary   102

7. Numerical Verifications for Ferronematics 105

   7.1   Solver Details 105
   7.2   Solutions of the Full Problem 106
   7.3   Solutions of the Reduced Problem 109
   7.4   Asymptotics Checking for $k \to \infty$ 112
   7.5   Bifurcation Diagrams 114
   7.6   Summary 118

Part 3: Smectic Liquid Crystals   119

8. A Mathematical Model of Smectics 121

   8.1   The de Gennes Model 122
   8.2   The Pevnyi–Selinger–Sluckin Model 124
   8.3   Our Proposed Model 126
       8.3.1   A unified framework 127
       8.3.2   Existence of minimisers 129
   8.4   Summary 133

9. Finite Element Discretisation 135

   9.1   Some Basics About Finite Elements 135
   9.2   *A Priori* Analysis for $q = 0$ 136
       9.2.1   *A priori* error estimates for $(\mathcal{P}1)$ 137
       9.2.2   *A priori* error estimates for $(\mathcal{P}2)$ 146
   9.3   Convergence Tests 170
       9.3.1   Test 1: On the unit square 170
       9.3.2   Test 2: On the unit disc 176
   9.4   Summary 178

10. Numerical Experiments for Smectics 179

   10.1   Implementation Details 179
   10.2   Scenario I: Defect Free 180
   10.3   Scenario II: Focal Conic Domains 184
   10.4   Scenario III: Oily Streaks 189
   10.5   Summary 192

Part 4: Conclusions and Perspectives   **193**

11. Summary and Conclusions   195

    11.1   Conclusions 195
    11.2   Potential Future Work 196
        11.2.1 Future work I 196
        11.2.2 Future work II 197
        11.2.3 Future work III 198

Appendix A   203

    A.1   Stability of the Newton System 203
    A.2   Equilibrium Equations in Two Dimensions 204

Bibliography   209

Index   219

List of Figures

1.1 Three types of molecular orientations in LC. As the temperature is increased, the material goes from solid or crystal through the cholesteric, smectic, nematic and liquid phases. 2

3.1 V-cycle; S represents smoothing, R restriction of residual and P prolongation. 39

3.2 W-cycle; S represents smoothing, R restriction of residual and P prolongation. 39

3.3 F-cycle; S represents smoothing, R restriction of residual and P prolongation. 39

3.4 Illustrations of the star patch of the center vertex (a) and the point-block patch (b) for the finite element pair $[\mathbb{P}_2]^2$-$\mathbb{P}_1$. Note that these two patches are the same for $[\mathbb{P}_1]^2$-$\mathbb{P}_1$ discretisation (c). Here, black dots represent the degrees of freedom, and the blue lines gather degrees of freedom solved for simultaneously in the relaxation. 45

4.1 A sample solution of the twist configuration. Colours represent the magnitude of directors. 56

4.2 Average number of FGMRES iterations per nonlinear iteration when continuing in K_2 for the LC problem in a square slab. 58

4.3 Average number of FGMRES iterations per nonlinear iteration when continuing in q_0 for the LC problem in a square slab. 58

4.4 The convergence of the computed director as the mesh is refined for the nematic LC problem in a square slab. 59

4.5   The coarse mesh of the ellipse.   60

4.6   Comparison of the computed constraint $\|\mathbf{n} \cdot \mathbf{n} - 1\|_0$ and the reference line $\mathcal{O}(\gamma^{-1/2})$ using the $[\mathbb{P}_1]^3$–$\mathbb{P}_1$ finite element pair for equal-constant nematic LC problems in an ellipse.   61

5.1   The profiles of $\mathbf{p}$ and their corresponding transition costs in (5.28).   80

6.1   Scalar order parameter s (a) under radial boundary conditions (6.5) and its corresponding director field (b). Note that $s \approx 0$ implies the presence of defects.   83

6.2   Plot of the scalar order parameter s on a unit square (a) and a rectangle (b).   84

6.3   A clip (along $y = 0$) of the biaxiality parameter β (a) and director (b). Notice that the point defect at the center has the value of $\beta \approx 0.35$, implying that this $+1$ defect is actually biaxial rather uniaxial.   86

6.4   Three-dimensional plot of the biaxiality parameter β (a) and director (b) in a unit cube. There is a plane defect at $z = 0.5$ having $\beta \approx 0$ and it is surrounded by the plane of $\beta \approx 0.98$. . . .   87

6.5   Results of nematic liquid crystals in a unit cube with planar radial boundary conditions (6.8): the biaxiality parameter β (a), a clip (along $x = 1/2$) of β (b) and the director on the plane of $z = 1$ (c). Notice that a point defect ($\beta \approx 0$) is presented in the rightmost picture; again, it is surrounded by points that have $\beta \approx 1$.   87

6.6   Aircraft stiffener profiles used in this work and their corresponding geometries. (a) L-shaped asymmetric stiffener; (b) L-shaped symmetric stiffener; (c) Z-shaped asymmetric stiffener.   96

6.7   (a) The bifurcation diagram of the L-shaped asymmetric stiffener where the functional $\mathbf{u}_1(0.019, 0, 0.025)$ corresponds to the y-component of the displacement evaluated at the midpoint of the left boundary. (b) The average stress at the bottom face. The enumeration of the branches from B1 to B7 corresponds to the images shown in Figure 6.8.   97

6.8   Seven buckling modes of the L-shaped asymmetric stiffener at $\varepsilon = 0.002$. The colours refer to the magnitude of the displacement from the original configuration.   98

6.9 (a) The bifurcation diagram of the L-shaped symmetric stiffener where the functional $\mathbf{u}_1(0.019, 0, 0.025)$ is the y-component of the displacement at the midpoint $(0.019, 0, 0.025)$ of the left boundary. (b) The average stress at the bottom face. The enumeration of the branches from B1 to B7 corresponds to the images shown in Figure 6.10. 99

6.10 Seven buckling modes of the L-shaped symmetric stiffener at $\varepsilon = 0.0007$. The colours refer to the magnitude of the displacement. 100

6.11 (a) The bifurcation diagram of the Z-shaped asymmetric stiffener where the functional $\mathbf{u}_0(0, 0.01635, 0.03)$ is taken to be the x-component of the displacement at the centre $(0, 0.01635, 0.03)$ of the flange. (b) The average stress at the bottom face. The enumeration of the branches from B1 to B9 corresponds to the images shown in Figure 6.12. 101

6.12 Nine buckling modes for the Z-shaped asymmetric stiffener at $\varepsilon = 0.0027$. The colours refer to the magnitude of the displacement. 102

7.1 Four stable stationary profiles, $(Q_{11}, Q_{12}, M_1, M_2)$, of (5.6) with $k = 0.01$ and $c = \xi = 1$, along with plots of $(2\phi - \theta)$, θ, and ϕ to verify the relation (5.18). Solutions 1 and 2 have the lowest full energy value (5.6). 107

7.2 Two examples of stable stationary profiles $(Q_{11}, Q_{12}, M_1, M_2)$ of the full energy (5.6) with $k = 0.01$, $c = 5$ and $\xi = 1$, along with plots of $(2\phi - \theta)$, θ, and ϕ to verify the relation (5.18). Solution 2 has lower energy than Solution 1. 108

7.3 The only (stable) solution of (5.19) for $c = \xi = 1$, and $k = 10$. 110

7.4 Four OR solution profiles with $c = \xi = 1$ and $k = 0.01$. Solution 1 is the OR energy minimiser (5.19). 111

7.5 Two stable OR critical points of (5.19), for $c = 5$, $\xi = 1$ and $k = 0.01$. The right profile has lower OR energy than the left profile and the solutions in Figure 7 6. 112

7.6 Two stable OR solutions with single (left) and multiple (right) interior transition layers for $c = 5$, $\xi = 1$ and $k = 0.01$. The left profile has lower OR energy than the right profile. 112

7.7   Log–log plots of $\|Q_{11}^{num} - Q_{11}^{asymp}\|_\infty$ (top row) and $\|M_1^{num} - M_1^{asymp}\|_\infty$ (bottom row). Left: truncating asymptotic expansions (7.3) and (7.4) at c^0. Middle: truncating asymptotic expansions at c^1. Right: truncating asymptotic expansions at c^2.   115

7.8   (a) The bifurcation diagram of continuing $k_1 = k_2 = k \in [0.2, 3.0]$ with fixed $c = \xi = 1$; here, black represents unstable solutions while blue indicates stable solutions. (b) The stable solution for $k = 2$.   115

7.9   Three solutions for $k = 1$ in Figure 7.8. Solutions 1 and 3 are global energy minimisers.   116

7.10   Eight new solutions for $k = 0.2$ in Figure 7.8. Solutions 1 and 8 are global energy minimisers.   117

7.11   (a) The bifurcation diagram with fixed $c = 5$ and $\xi = 1$; here, black labels unstable solutions while blue labels stable solutions. (b) One stable OR solution for $k = 4.45$.   118

7.12   Two new stable solutions at $k = 4.43$ in Figure 7.11.   118

8.1   Graphical illustrations of nematic, smectic-A and smectic-C phases. The top and bottom substrate plates are polarisers with perpendicular alignment directions. This type of polarisers is for example used in twisted nematic display (Dunmur and Sluckin, 2011, Technical Box 10.1).   122

10.1   The bifurcation diagram of the defect-free scenario.   182

10.2   Stationary states obtained at different values of θ_0 in the defect-free scenario. The visualisation displays the density perturbation u. For each solution, the value of the energy functional per unit area is displayed above it and we specify the stable profiles with asterisks. The bottom row depicts the lowest energy solution found for each value of θ_0.   183

10.3   (a) The first converged solution using Newton's method on a mesh of $6 \times 6 \times 5$ hexahedra using the TFCD settings. (b) Another solution profile with single screw dislocation around the central axis of the cuboid. The solution with screw dislocation has higher energy and both are stable. The gray layers are zero iso-surfaces of the density variation u.   186

10.4 Three numerical solutions for $\theta_c = \frac{\pi}{12}$ on a mesh of $6 \times 6 \times 5$ hexahedra. The solution with double screw dislocations has highest energy while the FCD solution possesses lowest energy. All profiles are stable. 187

10.5 Two solution profiles by taking $\theta_c = \frac{\pi}{10}$ on a mesh of $6 \times 6 \times 5$ hexahedra. The solution with screw dislocation has higher energy. Both profiles are stable. 188

10.6 Two numerical solutions for $\theta_c = \frac{\pi}{8}$ on a mesh of $6 \times 6 \times 5$ hexahedra. The solution with screw dislocation has higher energy. Both profiles are stable. 188

10.7 Eccentricity of FCD solutions as a function of preferred surface alignment angle. 189

10.8 Oily streaks. (a)–(c) Candidate structures proposed in Michel *et al.* [Lacaze *et al.* (2007)] consistent with X-ray diffraction. (d) Bifurcation diagram of structures as a function of aspect ratio. (e) Selected stationary states obtained at different aspect ratio r. The top row represents the lowest energy solution found. For each solution, the value of the energy functional per unit area is displayed below it with asterisks indicating stable profiles. 191

11.1 Four solution profiles and their stabilities with strongly-enforced Dirichlet data on $\delta\rho = 1$ and strongly-enforced homeotropic boundary conditions of $\mathbf{Q}$ on top and bottom surfaces of a wedge. Solution 4 with three edge dislocations has the lowest energy. 199

11.2 Meshes of two fused annuli. The domains differ in the sizes of the inclusions. 200

11.3 Two solution profiles of the geometry with inclusion ratio 0.2. 200

11.4 Two solution profiles for the geometry with inclusion ratio 0.4. 201

List of Tables

3.1 The average number of FGMRES iterations per Newton iteration (total number of Newton iterations) for a nematic LC problem in a square slab. See the detailed problem description in Chapter 4. 34

4.1 A comparison of the nonlinear convergence of the Newton linearisation (2.16) and the Picard iteration (2.17) using ideal inner solvers for a nematic LC problem in a square slab. The table shows the average number of outer FGMRES iterations per nonlinear iteration and the total nonlinear iterations in brackets. 56

4.2 ALLU: The average number of FGMRES iterations per nonlinear iteration for a nematic LC problem in a square slab using $[\mathbb{P}_2]^3$–$\mathbb{P}_1$ discretisation. Note here the last four columns are excerpted from Table 4.1 using the Picard iteration. 57

4.3 ALMG-STAR: The average number of FGMRES iterations per nonlinear iteration (total Newton iterations) for the nematic LC problem in a square slab. 57

4.4 ALMG-PBJ: The average number of FGMRES iterations per nonlinear iteration (total Newton iterations) for the nematic LC problem in a square slab. 58

4.5 The computing time of ALMG-PBJ, ALMG-STAR and MGVANKA as a function of mesh refinement for the nematic LC problem in a square slab. 60

4.6 ALLU: The average number of FGMRES iterations per nonlinear iteration for an equal-constant nematic problem in an ellipse using $[\mathbb{P}_2]^3$–$\mathbb{P}_1$ discretisation. 61

4.7   ALMG-STAR: The average number of FGMRES iterations per nonlinear iteration (total nonlinear iterations) for
equal-constant nematic problem in an ellipse using $[\mathbb{P}_2]^3$–$\mathbb{P}_1$
discretisation.   62
4.8   ALMG-PBJ: The average number of FGMRES iterations per
Newton iteration (total Newton iterations) for equal-constant
nematic problem in an ellipse using $[\mathbb{P}_2]^3$–$\mathbb{P}_1$ discretisation. . . .   62

9.1   Test 1: Convergence rates for $\mathbf{Q}$ with different degrees of polynomial approximation, in the decoupled case $q = 0$.   172
9.2   Test 1: Convergence rates using the consistent discrete formulation (9.30) with penalty parameter $\epsilon = 1$ and different polynomial degrees, in the decoupled case $q = 0$.   173
9.3   Test 1: Convergence rates using the inconsistent discrete formulation (9.66) with penalty parameter $\epsilon = 1$ and different
polynomial degrees, in the decoupled case $q = 0$.   174
9.4   Test 1: Convergence rates using the inconsistent discrete formulation (9.66) with penalty parameter $\epsilon = 5 \times 10^4$ and different
polynomial degrees, in the decoupled case $q = 0$.   174
9.5   Test 1: Convergence rates for u with $q = 30$ and penalty parameter $\epsilon = 5 \times 10^4$ with the inconsistent discretisation (9.66)
for u, fixing the approximation for $\mathbf{Q}$ to be with the $[\mathbb{Q}_2]^2$
element.   175
9.6   Test 1: Convergence rates for $\mathbf{Q}$ with $q = 30$ and penalty
parameter $\epsilon = 5 \times 10^4$ with the inconsistent discretisation
(9.66) for u, fixing the approximation for u to be with the $\mathbb{Q}_3$
element.   175
9.7   Test 2: Convergence rates for $\mathbf{Q}$ with different degrees of polynomial approximation, in the decoupled case $q = 0$.   177
9.8   Test 2: Convergence rates using the inconsistent discrete formulation (9.66) with penalty parameter $\epsilon = 1$ and different
polynomial degrees, in the decoupled case $q = 0$.   177

Chapter 1

Introduction

1.1 Background

There are three common known states of matter in nature: gases, liquids, and solids. Within certain range of temperature variations, one expects to see the transition among those states. Under some exceptional conditions, other states of matter can also exist. Liquid crystals (LCs) are one of such examples.

LC, first discovered by Reinitzer (1888), are materials that can exist in an intermediate mesophase between isotropic (i.e., the physical properties are uniform in all directions) liquids and solid crystals. That is to say, LC can flow like liquids while also possessing long-range orientational order. Based on different ordering symmetries, Friedel (1922) proposed to classify them into three broad categories: *nematic, smectic* and *cholesteric*, as shown in Figure 1.1. In the following, we briefly summarise the characteristics of these phases (refer to Chandrasekhar, 1992; Stewart, 2004; Wang *et al.*, 2021 for further details).

Nematic phase: This is the simplest and most extensively studied form of LC phase where the molecules are not layered but tend to point in the same direction. The molecules are free to move (rotate or slide) in this phase and align approximately parallel to each other, thus giving a long-range orientational ordering.

Smectic phase: The molecules have similar orientation and point in the same direction as the molecules in nematic LC do but they also tend to line up into layers. Depending on the angles formed between the molecular axes and the planes of molecules, there are a number of different smectic phases. Figure 1.1 depicts the simplest smectic structure, the so-called smectic-A phase.

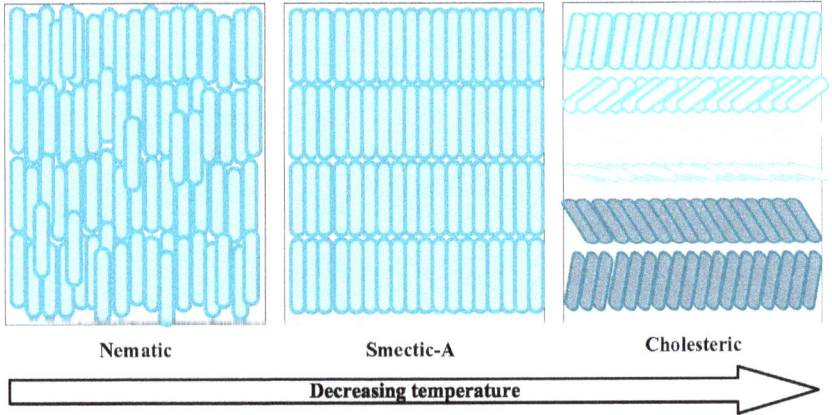

Fig. 1.1. Three types of molecular orientations in LC. As the temperature is increased, the material goes from solid or crystal through the cholesteric, smectic, nematic and liquid phases.

Note: For example, 8CB melts from crystal at 22°C to the smectic phase, then transitions to the nematic phase at 34°C and becomes a conventional liquid above 42°C (SciencebyDegrees, 2018).

Cholesteric phase: This is also known as chiral nematic phase and is characterised by the molecules being aligned and stacked in a helical pattern, with each layer rotated at an angle to the ones above and below it. It has a fixed pitch in its helical structure and is the last phase before the substance becomes a crystal or solid by decreasing the temperature.

Since the orientational properties of LC can be manipulated by imposing electric fields, they are often used to control light and have formed the basis of several important technologies in the area of electric display devices, e.g., digital screens. This has substantially increased interest in the scientific study of liquid crystals. Some examples of thorough overviews on LC modelling and its history can be found in Ball (2017), Chandrasekhar (1992), and Stewart (2004). More relevant to this book, there are two main continuum theories for modelling nematic LC: the Oseen–Frank and Landau–de Gennes theories, differing in the order parameters they use to describe the system. They both postulate a free energy, the minimisation of which gives the state of the LC. We include the detailed introduction of each theory later in the relevant part of this book.

The working flow of this book is as follows. We start with the director-based Oseen–Frank model for cholesteric and nematic liquid crystals in Part 1. The presence of the unit-length constraint on the director in this model stimulates the need for an efficient and robust solver for the saddle point systems arising in finite element discretisations of the equations. This is inspired by the work Benzi and Olshanskii (2006) and Farrell *et al.* (2019) for enforcing the divergence-free constraint in the stationary Navier–Stokes equations by applying the discrete augmented Lagrangian formulation. We propose an augmented-Lagrangian-type preconditioner and derive some robustness estimates in this part.

With this first experience of the Oseen–Frank model, its limitations in characterising more complicated defects (such as half-charge defects) become apparent, since it does not respect the head-to-tail symmetry of LC molecules. To explore and understand the typical defect structures, e.g., oily streaks (see Michel *et al.*, 2004, Figs. 9 and 16) and focal conics (see Williams and Kléman, 1975, Fig. 1) in smectics, we begin considering the Landau–de Gennes model employing a $\mathbf{Q}$-tensor as the state variable. As a step in this direction, Part 2 explores the effectiveness of the $\mathbf{Q}$-tensor model in characterising defects by considering a model problem of ferronematics, where magnetic nanoparticles (MNPs) are suspended in a nematic LC and thus induce a spontaneous magnetisation response without any external magnetic fields. In this part, we study the solution landscapes of the ferronematic problem for different parameter spaces and focus on the numerical validations of some theoretical analyses proven by Dalby and Majumdar in Dalby *et al.* (2022).

This substantial success in observing some interesting defect structures in ferronematics stimulates our interest in investigating more sophisticated defects in smectics and thereby leads to our work in Part 3. We propose a new mathematical model for smectic-A LC in this last part, which successfully captures typical structures of oily streaks and focal conic domains. We believe it can be applied to many other smectic scenarios that require an effective and efficient mathematical model.

Following this working flow, we divide the remainder of this book into three parts regarding different applications in liquid crystals, i.e., cholesterics (and nematics), ferronematics and smectics, and close with some conclusions and potential directions for future work. Each part majorly expands upon relevant publications, as detailed below.

Part 1   • *Xia, Farrell, and Wechsung* (2021) (Xia *et al.*, 2021a), published in *BIT Numerical Mathematics.*

Part 2   • *Dalby, Farrell, Majumdar, and Xia* (2022) (Dalby *et al.*, 2022), published in *SIAM Journal on Applied Mathematics.*

   • *Xia, Farrell, and Castro* (2020) (Xia *et al.*, 2020), published in *Thin-Walled Structures.*

Part 3   • *Xia, MacLachlan, Atherton, and Farrell* (2021) (Xia *et al.*, 2021b), published in *Physical Review Letters.*

   • *Xia and Farrell* (2023) (Xia and Farrell, 2023), published in *ESAIM: Mathematical Modelling and Numerical Analysis.*

1.2   Some Common Notation

d	spatial dimension, $d \in \{1, 2, 3\}$
Ω	open, bounded d-dimensional domain with Lipschitz boundary $\partial\Omega$
x, y, z	coordinates of domain Ω
C	generic constant that may not be the same constant for each appearance
h	mesh size
$\mathcal{T}_h$	mesh of Ω
T	element of $\mathcal{T}_h$
$\mathcal{E}_I$	set of all interior edges/faces of the mesh $\mathcal{T}$
$\mathcal{E}_B$	set of all boundary edges/faces of the mesh $\mathcal{T}$
$\mathcal{E}$	set of all edges/faces of the mesh $\mathcal{T}$; $\mathcal{E} = \mathcal{E}_I \cup \mathcal{E}_B$
S_0	space of symmetric, traceless $d \times d$ matrices
$\mathcal{S}^{d-1}$	($d \in \{2, 3\}$) surface of the unit ball in $\mathbb{R}^d$ centered at the origin
$\mathbb{M}^{d \times d}$	space of all $d \times d$ matrices
$\mathbf{I}_d$	identity matrix in $\mathbb{M}^{d \times d}$, $\mathbf{I}$ general identity matrix
$\mathcal{A}$	admissible space of a minimisation problem
$\mathbb{P}^{\Bbbk}$	piecewise continuous polynomials of degree $\Bbbk \geq 0$ on a simplicial mesh (intervals, triangles and tetrahedra)
$\mathbb{Q}^{\Bbbk}$	piecewise continuous polynomials of degree $\Bbbk \geq 0$ on a mesh of quadrilaterals or hexahedra
ν	outward unit normal to the boundary $\partial\Omega$
$H^{\Bbbk}$	Sobolev space of square-integrable functionals with square-integrable weak derivatives up to $\Bbbk$ order with standard $H^{\Bbbk}$-norm $\| \cdot \|_{\Bbbk}$ on Ω

$\mathbf{H}^{\Bbbk}$	vector-valued version of $H^{\Bbbk}$
$H_b^{\Bbbk}$, $\mathbf{H}_b^{\Bbbk}$	Sobolev spaces $H^{\Bbbk}$, $\mathbf{H}^{\Bbbk}$ with an addition of the trace
$\|\cdot\|_0$, $\|\cdot\|_\infty$	standard L^2- and L^∞-norm on Ω
$(\cdot,\cdot)_0$	inner product in $L^2(\Omega)$
$\Delta = \nabla \cdot \nabla$	Laplace operator
$\mathcal{D}^2$	Hessian operator

In order to avoid the proliferation of constants throughout this book, we use the notation $a \lesssim b$ (respectively, $b \gtrsim a$) to represent the relation $a \leq Cb$ (respectively, $b \geq Ca$) for some generic constant C independent of the mesh.

1.3 Common Solver Details

Since the problems to be solved in this book are all nonlinear, we always use Newton's method with L^2 linesearch (Brune *et al.*, 2015, Algorithm 2) as the outer nonlinear solver. The solver is implemented in the Firedrake (Rathgeber *et al.*, 2017) library, which relies on PETSc (Balay *et al.*, 2018) for solving the linear systems resulted from linearising the nonlinear problem. To install Firedrake, readers may find https://firedrakeproject.org/download.html quite useful.

In addition, for those problems (e.g., in Parts 2 and 3) where we are interested in multiple solutions or bifurcation diagrams, we use the so-called *deflation* technique as described in Section 6.2 and Farrell *et al.* (2015) to compute multiple solutions. This technique is implemented in the Defcon library (Farrell, 2017). See https://bitbucket.org/pefarrell/defcon/src/master/ for a detailed installation path.

Further details of each solver used for the numerical experiments in Chapters 4, 7, and 10 will be specified later. For reproducibility, the exact versions of the implementation codes used have been archived on Zenodo; the appropriate archived code will be cited in the corresponding chapter.

Part 1

Nematic and Cholesteric Liquid Crystals

Chapter 2

A Mathematical Model of Nematics and Cholesterics*

As mentioned in the previous chapter, one of the commonly used mathematical models for nematic and cholesteric liquid crystals is the Oseen–Frank theory (Frank, 1958; Oseen, 1933), which takes a unit-length vector field as the state variable. We therefore introduce the form of the Oseen–Frank model that we will subsequently consider.

2.1 The Oseen–Frank Model

Let $\Omega \subset \mathbb{R}^d, d \in \{2, 3\}$ be an open, bounded domain with Lipschitz boundary $\partial\Omega$. We triangulate the domain Ω and denote the mesh by $\mathcal{T}_h$ with each element represented by $\mathcal{T}$ and h is the mesh size. Denote $\mathbf{H}_b^1(\Omega) = \{\mathbf{v} \in H^1(\Omega, \mathbb{R}^3) : \mathbf{v}|_{\partial\Omega} = \mathbf{n}_b\}$ for a given vector field $\mathbf{n}_b \in H^{1/2}(\partial\Omega, \mathcal{S}^2)$ with $\mathbf{H}_0^1$ given by zero trace $\mathbf{n}_b = \mathbf{0} \notin \mathcal{S}^2$. Assume that the (nematic or) cholesteric LC occupying the domain Ω is equipped with a rigid anchoring (Dirichlet) boundary condition $\mathbf{n}|_{\partial\Omega} = \mathbf{n}_b$.[1] The Oseen–Frank model (Frank, 1958) considers the following minimisation problem:

$$\min_{\mathbf{n} \in \mathbf{H}_b^1(\Omega)} \mathcal{J}^{OF}(\mathbf{n}) = \int_\Omega W^{OF}(\mathbf{n}),$$
$$\text{subject to } \mathbf{n} \cdot \mathbf{n} = 1 \text{ a.e.,}$$
(2.1)

---

*This part of work majorly expands on the author's work (Xia *et al.*, 2021a).
[1]The following theory also applies with mixed periodic and Dirichlet boundary conditions (Adler *et al.*, 2015b; Bedford, 2014), which we shall use in some numerical examples in Chapter 4.

where the Frank energy density $W^{OF}(\mathbf{n})$ is of the form

$$W^{OF}(\mathbf{n}) = \frac{K_1}{2} (\nabla \cdot \mathbf{n})^2 + \frac{K_2}{2} (\mathbf{n} \cdot (\nabla \times \mathbf{n}) + q_0)^2 + \frac{K_3}{2} |\mathbf{n} \times (\nabla \times \mathbf{n})|^2$$

$$+ \frac{K_2 + K_4}{2} [\mathrm{tr}((\nabla \mathbf{n})^2) - (\nabla \cdot \mathbf{n})^2], \qquad (2.2)$$

with $\mathrm{tr}(\cdot)$ the trace of a matrix, $K_i \in \mathbb{R}$ ($i = 1, 2, 3, 4$) elastic constants (called *Frank constants*) and $q_0 \geq 0$ the preferred pitch for the cholesteric. K_1, K_2, K_3, and K_4 are referred to as the splay, twist, bend, and saddle-splay constants, respectively. Readers may refer to Kim *et al.* (2011, Figure 2) for a schematic illustration of these splay, twist and bend effects in liquid crystals. Note here $\nabla \mathbf{n}$ is matrix-valued and $(\nabla \mathbf{n})^2$ denotes the matrix multiplication of the matrix $\nabla \mathbf{n}$ and itself.

If $K_1 = K_2 = K_3 = K_c > 0$ and $K_4 = 0$, the energy density (2.2) reduces to the so-called *equal-constant* approximation, with energy density

$$W^{OF}(\mathbf{n}) = \frac{K_c}{2} \left[|\nabla \mathbf{n}|^2 + 2q_0 \mathbf{n} \cdot (\nabla \times \mathbf{n}) + q_0^2 \right],$$

which is a useful simplification to help us gain qualitative insight into more complex situations.

Remark 2.1. When $q_0 = 0$, the energy density (2.2) corresponds to the nematic case. Furthermore, when combined with the equal-constant approximation, (2.2) reduces to

$$W^{OF}(\mathbf{n}) = \frac{K_c}{2} |\nabla \mathbf{n}|^2.$$

With this free energy density, the solution to the minimisation problem (2.1) is unique and is known as the *harmonic map* from a two- or three-dimensional compact manifold to $\mathcal{S}^2$ (Lin, 1989). Some fast numerical algorithms for this equal-constant approximation case have been proposed and tested in Hu *et al.* (2009).

Using the fact that

$$\mathrm{tr}((\nabla \mathbf{n})^2) - (\nabla \cdot \mathbf{n})^2 = \nabla \cdot ((\mathbf{n} \cdot \nabla)\mathbf{n} - (\nabla \cdot \mathbf{n})\mathbf{n}),$$

the last term (the *saddle-splay* term or the *null Lagrangian*) in (2.2) can be dropped as its integral reduces to a surface integral, which is essentially a constant if applying Dirichlet boundary conditions to the model, via the divergence theorem. For mixed periodic and Dirichlet boundary conditions considered in Section 4.2.1, we can verify directly that this saddle-splay energy vanishes. Hence, for simplicity, it suffices to consider the following Frank energy density

$$W^{OF}(\mathbf{n}) = \frac{K_1}{2} (\nabla \cdot \mathbf{n})^2 + \frac{K_2}{2} (\mathbf{n} \cdot (\nabla \times \mathbf{n}) + q_0)^2 + \frac{K_3}{2} |\mathbf{n} \times (\nabla \times \mathbf{n})|^2.$$
$$(2.3)$$

In this chapter, we use a more compact form of the free energy (2.1) as in Adler *et al.* (2015b, 2016) by introducing a symmetric dimensionless tensor

$$\mathbf{Z} = \kappa \mathbf{n} \otimes \mathbf{n} + (\mathbf{I}_3 - \mathbf{n} \otimes \mathbf{n}) = \mathbf{I}_3 + (\kappa - 1)\mathbf{n} \otimes \mathbf{n},$$

where $\kappa = K_2/K_3$. By the classical equality

$$|\nabla \times \mathbf{n}|^2 = (\mathbf{n} \cdot (\nabla \times \mathbf{n}))^2 + |\mathbf{n} \times (\nabla \times \mathbf{n})|^2, \qquad (2.4)$$

the original energy functional $\mathcal{J}^{OF}(\mathbf{n})$ can be rewritten as

$$\mathcal{J}^{OF}(\mathbf{n}) = \frac{1}{2} (K_1 (\nabla \cdot \mathbf{n}, \nabla \cdot \mathbf{n})_0 + K_3 (\mathbf{Z} \nabla \times \mathbf{n}, \nabla \times \mathbf{n})_0$$
$$+ 2K_2 q_0 (\mathbf{n}, \nabla \times \mathbf{n})_0 + K_2 (q_0, q_0)_0). \qquad (2.5)$$

It can be observed that the auxiliary tensor $\mathbf{Z}$ contributes to the nonlinearity of $\mathcal{J}^{OF}(\mathbf{n})$ in (2.5).

Remark 2.2. There is another widely used simplification of the energy density (2.2), where $q_0 = 0$ and $K_2 = K_3 = K_1 + K_p$, $K_4 = -K_p$ (Glowinski *et al.*, 2003; Lin and Richter, 2007). In this case, (2.2) becomes

$$W^{OF}(\mathbf{n}) = \frac{1}{2} [K_1 |\nabla \mathbf{n}|^2 + K_p |\nabla \times \mathbf{n}|^2],$$

and it is expected that as $K_p \to \infty$, the asymptotic behavior of minimisers provides a description of the phase transition process of LC from the nematic phase to the smectic-A phase (Glowinski *et al.*, 2003; Lin and Richter, 2007; Lin and Tai, 2014).

Furthermore, it is proven in Adler *et al.* (2015b, Section 2.3) that $\mathbf{Z}$ is uniformly (with respect to $\mathbf{x} \in \Omega$) symmetric positive definite (USPD) as long as sufficient control is maintained on $|\mathbf{n}|^2$. This property of $\mathbf{Z}$ plays an essential role in proving the well-posedness of the saddle-point problem in the nematic case. We restate the result of $\mathbf{Z}$ being USPD in the following, as it is important later:

Lemma 2.1 (Adler *et al.*, 2015b, Section 2.3). *Assume $\alpha \leq |\mathbf{n}|^2 \leq \beta \ \forall \mathbf{x} \in \Omega$ with $0 < \alpha \leq 1 \leq \beta$. If $\kappa > 1$, then $\mathbf{Z}$ is USPD on Ω; for $0 < \kappa < 1$, then $\mathbf{Z}$ is USPD on Ω if $\beta < \frac{1}{1-\kappa}$.*

Remark 2.3. Notice that the regularity of $\mathbf{n} \in \mathbf{H}^1(\Omega)$ is enough for the functional $\mathcal{J}^{OF}(\mathbf{n})$ of (2.5) to be well defined. In fact, $\mathbf{n} \in \mathbf{H}^1(\Omega)$ implies $\nabla \cdot \mathbf{n} \in L^2(\Omega)$ and $\nabla \times \mathbf{n} \in \mathbf{L}^2(\Omega)$. By (2.4), we have $\mathbf{n} \cdot (\nabla \times \mathbf{n}) \in L^2(\Omega)$. This ensures that the term $(q_0, \mathbf{n} \cdot (\nabla \times \mathbf{n}))_0$ in (2.5) is defined. Furthermore, Lemma 2.1 gives the boundedness of $\mathbf{Z}$, which guarantees the L^2-regularity of the term $\mathbf{Z}\nabla \times \mathbf{n}$ in (2.5).

Naturally, the values of elastic constants and the cholesteric pitch will be an important factor in determining the minimisers. In particular, the free energy density should be bounded from below so to ensure the existence of minimisers. With an addition of arbitrary constant, we thus need additional assumptions on the parameters to satisfy non-negativity of the energy density, i.e.,

$$W^{OF}(\mathbf{n}) \geq 0 \quad \forall \mathbf{n} \in \mathbf{H}_b^1(\Omega).$$

This gives rise to Ericksen's inequalities (see Ball, 2017; Bedford, 2014 and references therein):

$$K_1, K_2, K_3 \geq 0, K_2 + K_4 = 0 \qquad \text{if } q_0 \neq 0,$$
$$2K_1 \geq K_2 + K_4, K_2 \geq |K_4|, K_3 \geq 0 \quad \text{if } q_0 = 0.$$

Remark 2.4. We have included the inequalities with regard to constant K_4 here for generality, though they are not necessary in our work as we have eliminated the K_4-related term in the free energy. In this part, we will simply consider $K_i > 0$ ($i = 1, 2, 3$) to avoid any technical issues.

For the minimisation problem (2.1) arising in (nematic or cholesteric) liquid crystals, it has been proven in Lin (1989, Theorem 2.1) that there exists a solution.

Theorem 2.2 (Lin, 1989, Theorem 2.1). *Let Ω be a bounded Lipschitz domain and assume the Dirichlet boundary data $\mathbf{n}_b \in H^{1/2}(\partial\Omega, \mathcal{S}^2)$. If $K_1, K_2, K_3 > 0$, then there exists an $\mathbf{n} \in H_b^1(\Omega, \mathcal{S}^2) := \{\mathbf{n} \in H^1(\Omega, \mathcal{S}^2) : \mathbf{n} = \mathbf{n}_b \text{ on } \partial\Omega\}$ such that*

$$\mathcal{J}^{OF}(\mathbf{n}) = \inf_{\mathbf{u} \in H_b^1(\Omega, \mathcal{S}^2)} \mathcal{J}^{OF}(\mathbf{u}).$$

The main difficulty in numerically solving the Oseen–Frank model (2.1) is the enforcement of the unit-length constraint. There are several existing approaches to handling constraints, e.g., projection (Lin and Tai, 2014), Lagrange multipliers, and penalty methods (Nocedal and Wright, 1999, Section 12.3 and 17).

The projection method is numerically simple but the value of the energy functional may go up and down dramatically after each projection, making it difficult to control in the optimisation procedure (Lin and Tai, 2014). A Lagrange multiplier is often used to replace constrained optimisation problems with unconstrained ones, but an important disadvantage of this approach is that it introduces another unknown (i.e., the Lagrange multiplier) and leads to a saddle-point structure which can be difficult to solve (Benzi *et al.*, 2005). On the other hand, the penalty method has the favorable property that the resulting system has an energy decay property (Lin and Richter, 2007) which may result in an easier theoretical and numerical study of the solution. However, the penalty parameter has to be very large for the accuracy of approximating the constraints, leading to an ill-conditioned system. Some works based on either projection or pure penalty methods for nematic phases can be found in Glowinski and Le Tallec (1989), Glowinski *et al.* (2003), Lin and Richter (2007) and the references therein.

Fortunately, it is possible to amend the ill-conditioning effects with large penalty parameters that are inherent in the pure penalty method by combining it with a Lagrange multiplier. This is the *augmented Lagrangian* algorithm (Fortin and Glowinski, 1983). This strategy combines the advantages of both schemes: the penalty parameter can be relatively small due to the presence of the Lagrange multiplier, and the Schur complement of the saddle-point system is easier to solve due to the presence of the penalty term (Benzi and Olshanskii, 2006; Farrell *et al.*, 2019; Glowinski and Le Tallec, 1989; Glowinski *et al.*, 2003; Olshanskii, 2002). Since the concept of the Schur complement is closely related to this part of the book, we briefly summarise the approach of Schur complement reduction here. Consider a saddle-point system (that is, it has both positive and negative eigenvalues) of form

$$\mathbf{D} \begin{bmatrix} \mathbf{x} \\ \mathbf{y} \end{bmatrix} := \begin{bmatrix} \mathbf{A}_1 & \mathbf{B}_1^\top \\ \mathbf{B}_2 & \mathbf{C}_1 \end{bmatrix} \begin{bmatrix} \mathbf{x} \\ \mathbf{y} \end{bmatrix} = \begin{bmatrix} \mathbf{c} \\ \mathbf{d} \end{bmatrix}. \tag{2.6}$$

Assuming that both $\mathbf{A}_1$ and $\mathbf{D}$ are nonsingular, it implies that $\mathbf{S}_1 = \mathbf{C}_1 - \mathbf{B}_2 \mathbf{A}_1^{-1} \mathbf{B}_1^\top$ is also nonsingular (Benzi *et al.*, 2005). Here, $\mathbf{S}_1$ is the so-called *Schur complement*. Block Gaussian elimination then reduces the system

(2.6) to

$$\begin{bmatrix} \mathbf{A}_1 & \mathbf{B}_1^\top \\ \mathbf{0} & \mathbf{S}_1 \end{bmatrix} \begin{bmatrix} \mathbf{x} \\ \mathbf{y} \end{bmatrix} = \begin{bmatrix} \mathbf{c} \\ \mathbf{d} - \mathbf{B}_2\mathbf{A}_1^{-1}\mathbf{c} \end{bmatrix}. \tag{2.7}$$

If it is possible to solve linear systems involving $\mathbf{A}_1$ and $\mathbf{S}_1$, we can solve the coupled linear system.

In what follows, we first consider the method of Lagrange multipliers and then add the augmented Lagrangian term to control the Schur complement of the system.

2.2 Lagrange Multiplier and Newton Linearisation

By introducing the Lagrange multiplier $\lambda \in L^2(\Omega)$, the associated Lagrangian of the minimisation problem (2.1) is then defined as

$$\mathcal{L}(\mathbf{n}, \lambda) = \mathcal{J}^{OF}(\mathbf{n}) + (\lambda, \mathbf{n} \cdot \mathbf{n} - 1)_0, \tag{2.8}$$

and its first-order optimality conditions are: find $(\mathbf{n}, \lambda) \in \mathbf{H}_b^1(\Omega) \times L^2(\Omega)$ such that

$$\begin{aligned}
\mathcal{L}_{\mathbf{n}}[\mathbf{v}] &= \mathcal{J}_{\mathbf{n}}^{OF}[\mathbf{v}] + (\lambda, 2\mathbf{n} \cdot \mathbf{v})_0 \\
&= K_1 \left(\nabla \cdot \mathbf{n}, \nabla \cdot \mathbf{v}\right)_0 + K_3 \left(\mathbf{Z}\nabla \times \mathbf{n}, \nabla \times \mathbf{v}\right)_0 \\
&\quad + (K_2 - K_3)\left(\mathbf{n} \cdot \nabla \times \mathbf{n}, \mathbf{v} \cdot \nabla \times \mathbf{n}\right)_0 \\
&\quad + K_2 q_0 \left(\mathbf{v}, \nabla \times \mathbf{n}\right)_0 + K_2 q_0 \left(\mathbf{n}, \nabla \times \mathbf{v}\right)_0 + (\lambda, 2\mathbf{n} \cdot \mathbf{v})_0 \\
&= 0 \quad \forall \mathbf{v} \in \mathbf{H}_0^1(\Omega), \\
\mathcal{L}_{\lambda}[\mu] &= (\mu, \mathbf{n} \cdot \mathbf{n} - 1)_0 = 0 \quad \forall \mu \in L^2(\Omega).
\end{aligned} \tag{2.9}$$

As (2.9) is nonlinear, Newton linearisation is employed. Let $\mathbf{n}_j$ and λ_j be the current approximations for $\mathbf{n}$ and λ, respectively, and denote the corresponding updates to these approximations as $\delta \mathbf{n} = \mathbf{n}_{j+1} - \mathbf{n}_j$ and $\delta \lambda = \lambda_{j+1} - \lambda_j$. Then the Newton iteration at $(\mathbf{n}_j, \lambda_j)$ in block form is given by: find $(\delta \mathbf{n}, \delta \lambda) \in \mathbf{H}_0^1(\Omega) \times L^2(\Omega)$ such that

$$\begin{bmatrix} \mathcal{L}_{\mathbf{nn}} & \mathcal{L}_{\mathbf{n}\lambda} \\ \mathcal{L}_{\lambda\mathbf{n}} & 0 \end{bmatrix} \begin{bmatrix} \delta \mathbf{n} \\ \delta \lambda \end{bmatrix} = - \begin{bmatrix} \mathcal{L}_{\mathbf{n}} \\ \mathcal{L}_{\lambda} \end{bmatrix}, \tag{2.10}$$

where

$$
\begin{aligned}
\mathcal{L}_{\mathbf{nn}}[\mathbf{v}, \delta\mathbf{n}] &= J_{\mathbf{nn}}[\mathbf{v}, \delta\mathbf{n}] + (\lambda_j, 2\delta\mathbf{n} \cdot \mathbf{v})_0 \\
&= K_1 \left(\nabla \cdot \delta\mathbf{n}, \nabla \cdot \mathbf{v} \right)_0 + K_3 \left(\mathbf{Z}(\mathbf{n}_j) \nabla \times \delta\mathbf{n}, \nabla \times \mathbf{v} \right)_0 \\
&\quad + (K_2 - K_3) \Big((\delta\mathbf{n} \cdot \nabla \times \mathbf{n}_j, \mathbf{n}_j \cdot \nabla \times \mathbf{v})_0 \\
&\quad + (\mathbf{n}_j \cdot \nabla \times \mathbf{n}_j, \delta\mathbf{n} \cdot \nabla \times \mathbf{v})_0 \\
&\quad + (\mathbf{v} \cdot \nabla \times \mathbf{n}_j, \mathbf{n}_j \cdot \nabla \times \delta\mathbf{n})_0 + (\mathbf{n}_j \cdot \nabla \times \mathbf{n}_j, \mathbf{v} \cdot \nabla \times \delta\mathbf{n})_0 \\
&\quad + (\delta\mathbf{n} \cdot \nabla \times \mathbf{n}_j, \mathbf{v} \cdot \nabla \times \mathbf{n}_j)_0 \Big) \\
&\quad + K_2 q_0 \left(\mathbf{v}, \nabla \times \delta\mathbf{n} \right)_0 + K_2 q_0 \left(\delta\mathbf{n}, \nabla \times \mathbf{v} \right)_0 + (\lambda_j, 2\delta\mathbf{n} \cdot \mathbf{v})_0 ,
\end{aligned}
\tag{2.11}
$$

and

$$
\mathcal{L}_{\mathbf{n}\lambda}[\mathbf{v}, \delta\lambda] = (\delta\lambda, 2\mathbf{n}_j \cdot \mathbf{v})_0 ,
$$
$$
\mathcal{L}_{\lambda\mathbf{n}}[\mu, \delta\mathbf{n}] = (\mu, 2\mathbf{n}_j \cdot \delta\mathbf{n})_0 .
$$

Since $\mathcal{L}(\mathbf{n}, \lambda)$ is linear in λ, $\mathcal{L}_{\lambda\lambda} = 0$. This results in (2.10) being a saddle-point problem.

With a suitable spatial discretisation (we only consider conforming finite elements throughout this part of the book, i.e., the finite dimensional space $V_h \subset \mathbf{H}_0^1(\Omega)$ that the finite element approximation $\mathbf{n}_h$ of $\mathbf{n}$ belongs to, and the finite-dimensional space $Q_h \subset L^2(\Omega)$ that the approximation λ_h of λ belongs to), a symmetric saddle-point system must be solved at each Newton iteration:

$$
\begin{bmatrix} \mathbf{A} & \mathbf{B}^\top \\ \mathbf{B} & 0 \end{bmatrix} \begin{bmatrix} U \\ X \end{bmatrix} - \begin{bmatrix} \mathbf{f} \\ \mathbf{g} \end{bmatrix} ,
\tag{2.12}
$$

where U and X represent the coefficient vectors of $\delta\mathbf{n}$ and $\delta\lambda$ in terms of the basis functions of V_h and Q_h, respectively.

We can accordingly write the discrete variational problem as: find $\delta\mathbf{n}_h \in V_h$ and $\delta\lambda_h \in Q_h$ such that

$$
\begin{aligned}
\mathfrak{a}(\delta\mathbf{n}_h, \mathbf{v}_h) + \mathfrak{b}(\mathbf{v}_h, \delta\lambda_h) &= \mathfrak{f}(\mathbf{v}_h) \quad \forall \mathbf{v}_h \in V_h, \\
\mathfrak{b}(\delta\mathbf{n}_h, \mu_h) &= \mathfrak{g}(\mu_h) \quad \forall \mu_h \in Q_h,
\end{aligned}
\tag{2.13}
$$

where $\mathfrak{a}(\cdot,\cdot)$ and $\mathfrak{b}(\cdot,\cdot)$ are bilinear forms given by

$$\mathfrak{a}(\mathbf{u},\mathbf{v}) = K_1 \left(\nabla \cdot \mathbf{u}, \nabla \cdot \mathbf{v}\right)_0 + K_3 \left(\mathbf{Z}(\mathbf{n}_j)\nabla \times \mathbf{u}, \nabla \times \mathbf{v}\right)_0$$

$$+ (K_2 - K_3)\Big((\mathbf{u}\cdot\nabla\times\mathbf{n}_j, \mathbf{n}_j\cdot\nabla\times\mathbf{v})_0 + (\mathbf{n}_j\cdot\nabla\times\mathbf{n}_j, \mathbf{u}\cdot\nabla\times\mathbf{v})_0$$

$$+ (\mathbf{v}\cdot\nabla\times\mathbf{n}_j, \mathbf{n}_j\cdot\nabla\times\mathbf{u})_0 + (\mathbf{n}_j\cdot\nabla\times\mathbf{n}_j, \mathbf{v}\cdot\nabla\times\mathbf{u})_0$$

$$+ (\mathbf{u}\cdot\nabla\times\mathbf{n}_j, \mathbf{v}\cdot\nabla\times\mathbf{n}_j)_0 \Big)$$

$$+ K_2 q_0 \left(\mathbf{v}, \nabla\times\mathbf{u}\right)_0 + K_2 q_0 \left(\mathbf{u}, \nabla\times\mathbf{v}\right)_0 + (\lambda_j, 2\mathbf{u}\cdot\mathbf{v})_0,$$

and

$$\mathfrak{b}(\mathbf{v}, p) = (p, 2\mathbf{n}_j \cdot \mathbf{v})_0,$$

and $\mathfrak{f}$ and $\mathfrak{g}$ are linear functionals given by

$$\mathfrak{f}(\mathbf{v}) = -\Big(K_1 \left(\nabla\cdot\mathbf{n}_j, \nabla\cdot\mathbf{v}\right)_0 + K_3 \left(Z(\mathbf{n}_j)\nabla\times\mathbf{n}_j, \nabla\times\mathbf{v}\right)_0$$

$$+ (K_2 - K_3)\left(\mathbf{n}_j\cdot\nabla\times\mathbf{n}_j, \mathbf{v}\cdot\nabla\cdot\mathbf{n}_j\right)_0$$

$$+ K_2 q_0 \left(\mathbf{v}, \nabla\times\mathbf{n}_j\right)_0 + K_2 q_0 \left(\mathbf{n}_j, \nabla\times\mathbf{v}\right)_0$$

$$+ (\lambda_j, 2\mathbf{n}_j\cdot\mathbf{v})_0 \Big),$$

and

$$\mathfrak{g}(\mu) = -(\mu, \mathbf{n}_j\cdot\mathbf{n}_j - 1)_0.$$

Remark 2.5. The well-posedness of the continuous and discretised Newton system (with the $([\mathbb{Q}_\Bbbk]^d \oplus \mathbb{B}_F)$-$\mathbb{Q}_0$ finite element pair, $\Bbbk \geq 1$) for a generalised nematic LC problem is discussed in Adler *et al.* (2015b), where $\mathbb{B}_F := \{\mathbf{v} \in [\mathcal{C}_c(\Omega)]^d : \mathbf{v}|_T = a_T b_T \mathbf{n}_j|T \ \forall T \in \mathcal{T}_h\}$ denotes the bubble space. Here, $\mathcal{C}_c(\Omega)$ includes compactly supported continuous functions, b_T represents biquadratic bubble function that vanishes on $\partial T \in \mathcal{T}_h$ and satisfies

$$\begin{cases} \int_T b_T = 1 & \forall T \in \mathcal{T}_h, \\ b_T(x) > 0 & \forall x \in T, \end{cases}$$

and a_T is a constant associated with b_T. Moreover, the authors of Adler *et al.* (2016) considered the pure penalty approach for nematic LC and obtained a well-posedness result of the penalised Newton iteration through similar techniques. We will follow these analysis strategies in this section.

It is straightforward to deduce the well-posedness of the discrete Newton iteration (2.13) for cholesteric problems under some proper assumptions on the problem-dependent constants. In fact, two additional q_0-related terms in $\mathcal{L}_{nn}$ from (2.11) compared to the nematic energy density from Adler *et al.* (2015b) are simply L^2 inner products, which can be easily bounded above using the Cauchy–Schwarz and triangle inequalities. We start with the assumptions and subsequently prove some necessary ingredients, e.g., the coercivity and boundedness of $\mathfrak{a}(\cdot,\cdot)$ and the discrete inf-sup condition for $\mathfrak{b}(\cdot,\cdot)$, of the well-posedness result.

Assumption 2.3. *Assume that there exist constants $0 < \alpha \leq 1 \leq \beta$ such that $\alpha \leq |\mathbf{n}_j|^2 \leq \beta$. For $0 < \kappa < 1$, assume further that $\beta < \frac{1}{1-\kappa}$. By Lemma 2.1, $\mathbf{Z}(\mathbf{n}_j)$ remains USPD with lower bound Λ_l and upper bound Λ_u, i.e.,*

$$\Lambda_l \leq \frac{\mathbf{x}^\top \mathbf{Z}(\mathbf{n}_j)\mathbf{x}}{\mathbf{x}^\top \mathbf{x}} \leq \Lambda_u \quad \forall \mathbf{x} \in \mathbb{R}^d \backslash \{\mathbf{0}\}.$$

Lemma 2.4 (Continuous coercivity). *With Assumption 2.3, we assume further that the current Lagrange multiplier approximation λ_j is pointwise non-negative almost everywhere. Let $K_1 > K_2 q_0 C_4$ and $K_3 \Lambda_l > K_2 q_0 (C_4 + 1)$ with C_4 to be defined. Then there exists an $\alpha_0 > 0$ such that*

$$\alpha_0 \|\mathbf{v}\|_1^2 \leq \mathfrak{a}(\mathbf{v}, \mathbf{v}) \quad \forall \mathbf{v} \in \mathbf{H}_0^1(\Omega). \tag{2.14}$$

Moreover, when $\kappa = 1$, i.e., $K_2 = K_3$, if $K_1 > K_2 q_0 C_4$ and $1 > q_0(C_4 + 1)$, then the coercivity result (2.14) also holds.

Remark 2.6. One may wonder how realistic that λ_j can be pointwise non-negative almost everywhere during each nonlinear iteration. However, we do not observe any ill-posed problems during our numerical experiments that are illustrated in Chapter 4.

Proof. With the lower bound Λ_l of $\mathbf{Z}$, we compute:

$$\mathfrak{a}(\mathbf{v}, \mathbf{v}) \geq K_1 \|\nabla \cdot \mathbf{v}\|_0^2 + K_3 \Lambda_l \|\nabla \times \mathbf{v}\|_0^2 + 2K_2 q_0 \, (\mathbf{v}, \nabla \times \mathbf{v})_0 + 2\,(\lambda_j, \mathbf{v} \cdot \mathbf{v})_0$$

$$\geq K_1 \|\nabla \cdot \mathbf{v}\|_0^2 + K_3 \Lambda_l \|\nabla \times \mathbf{v}\|_0^2 - 2K_2 q_0 |\,(\mathbf{v}, \nabla \times \mathbf{v})_0\,|$$

$$\geq K_1 \|\nabla \cdot \mathbf{v}\|_0^2 + K_3 \Lambda_l \|\nabla \times \mathbf{v}\|_0^2 - 2K_2 q_0 \|\mathbf{v}\|_0 \|\nabla \times \mathbf{v}\|_0$$

$$\geq K_1 \|\nabla \cdot \mathbf{v}\|_0^2 + K_3 \Lambda_l \|\nabla \times \mathbf{v}\|_0^2 - K_2 q_0 (\|\mathbf{v}\|_0^2 + \|\nabla \times \mathbf{v}\|_0^2),$$

where the first inequality comes from the assumption that λ_j is non-negative pointwise and the last two inequalities are derived by Cauchy–Schwarz and Hölder inequalities, respectively.

By Girault and Raviart (2011, Remark 2.7), for a bounded Lipschitz domain, there exists $C_1 > 0$ such that

$$\|\nabla \mathbf{v}\|_0^2 \leq C_1(\|\nabla \cdot \mathbf{v}\|_0^2 + \|\nabla \times \mathbf{v}\|_0^2),$$

for all $\mathbf{v} \in \mathbf{H}_0(\text{div}, \Omega) \cap \mathbf{H}_0(\text{curl}, \Omega).$[2] Here, we denote

$$\mathbf{H}_0(\text{div}, \Omega) = \{\mathbf{v} \in \mathbf{L}^2(\Omega) : \nabla \cdot \mathbf{v} \in L^2(\Omega), \nu \cdot \mathbf{v} = 0 \text{ on } \partial\Omega\},$$
$$\mathbf{H}_0(\text{curl}, \Omega) = \{\mathbf{v} \in \mathbf{L}^2(\Omega) : \nabla \times \mathbf{v} \in \mathbf{L}^2(\Omega), \nu \times \mathbf{v} = \mathbf{0} \text{ on } \partial\Omega\}.$$

Then using the classical Poincaré inequality, $\|\mathbf{v}\|_0^2 \leq C_3\|\nabla\mathbf{v}\|_0^2$ for all $\mathbf{v} \in \mathbf{H}_0^1(\Omega)$, and defining $C_4 = C_1 C_3 > 0$, we have

$$\|\mathbf{v}\|_0^2 \leq C_4(\|\nabla \cdot \mathbf{v}\|_0^2 + \|\nabla \times \mathbf{v}\|_0^2).$$

Furthermore, there exists $C_2 = C_4 + C_1 > 0$ such that

$$\|\mathbf{v}\|_1^2 \leq C_2(\|\nabla \cdot \mathbf{v}\|_0^2 + \|\nabla \times \mathbf{v}\|_0^2).$$

It follows that

$$\begin{aligned}
\mathfrak{a}(\mathbf{v}, \mathbf{v}) &\geq K_1\|\nabla \cdot \mathbf{v}\|_0^2 + K_3\Lambda_l\|\nabla \times \mathbf{v}\|_0^2 \\
&\quad - K_2 q_0 \left[C_4 \left(\|\nabla \cdot \mathbf{v}\|_0^2 + \|\nabla \times \mathbf{v}\|_0^2\right) - \|\nabla \times \mathbf{v}\|_0^2\right] \\
&= (K_1 - K_2 q_0 C_4)\|\nabla \cdot \mathbf{v}\|_0^2 + (K_3\Lambda_l - K_2 q_0 C_4 - K_2 q_0)\|\nabla \times \mathbf{v}\|_0^2.
\end{aligned}$$

Choosing $C_5 = \min\{K_1 - K_2 q_0 C_4, K_3\Lambda_l - K_2 q_0 C_4 - K_2 q_0\} > 0$ (the positivity follows from the assumptions) and $\alpha_0 = C_5/C_2$, we find that the coercivity (2.14) holds.

In particular, when $\kappa = 1$ (i.e., $K_2 = K_3$), we have $\mathbf{Z} = \mathbf{I}_3$ and thus $\Lambda_l = 1$. Then, the bilinear form becomes

$$\begin{aligned}
\mathfrak{a}(\mathbf{v}, \mathbf{v}) &= K_1\|\nabla \cdot \mathbf{v}\|_0^2 + K_2\|\nabla \times \mathbf{v}\|_0^2 + 2K_2 q_0 \left(\mathbf{v}, \nabla \times \mathbf{v}\right)_0 + 2\left(\lambda_j, \mathbf{v} \cdot \mathbf{v}\right)_0 \\
&\geq K_1\|\nabla \cdot \mathbf{v}\|_0^2 + K_2\|\nabla \times \mathbf{v}\|_0^2 - 2K_2 q_0|\left(\mathbf{v}, \nabla \times \mathbf{v}\right)_0| \\
&\geq K_1\|\nabla \cdot \mathbf{v}\|_0^2 + K_2\|\nabla \times \mathbf{v}\|_0^2 - 2K_2 q_0\|\mathbf{v}\|_0\|\nabla \times \mathbf{v}\|_0 \\
&\geq K_1\|\nabla \cdot \mathbf{v}\|_0^2 + K_2\|\nabla \times \mathbf{v}\|_0^2 - K_2 q_0(\|\mathbf{v}\|_0^2 + \|\nabla \times \mathbf{v}\|_0^2).
\end{aligned}$$

By choosing $C_6 = \min\{K_1 - K_2 q_0 C_4, K_2(1 - q_0 C_4 - q_0)\} > 0$ (the positivity comes from the assumptions) and $\alpha_0 = C_6/C_2$, we obtain the desired

---

[2]In fact, $\mathbf{H}_0^1(\Omega) = \mathbf{H}_0(\text{div}, \Omega) \cap \mathbf{H}_0(\text{curl}, \Omega)$ holds for any bounded Lipschitz domain Ω (Girault and Raviart, 2011, Lemma 2.5).

coercivity

$$\mathfrak{a}(\mathbf{v}, \mathbf{v}) \geq \alpha_0 \|\mathbf{v}\|_1^2 \quad \forall \mathbf{v} \in \mathbf{H}_0^1(\Omega),$$

as stated in (2.14). □

So far, the coercivity of the bilinear form $\mathfrak{a}(\cdot, \cdot)$ has been shown for all functions in $\mathbf{H}_0^1(\Omega)$. Discrete coercivity follows if a conforming finite element for the director space is chosen.

The boundedness of the bilinear form $\mathfrak{a}(\cdot, \cdot)$ and the right-hand side functionals $\mathfrak{f}(\cdot)$ and $\mathfrak{g}(\cdot)$ can be obtained directly by following the proofs in Adler *et al.* (2015b). Hence, we omit the details here.

It remains to consider the discrete inf-sup condition of the bilinear form $\mathfrak{b}(\cdot, \cdot)$ for a finite element pair V_h-Q_h, i.e., whether there exists a constant C such that

$$\sup_{\mathbf{u}_h \in V_h \backslash \{0\}} \frac{\mathfrak{b}(\mathbf{u}_h, \mu_h)}{\|\mathbf{u}_h\|} \geq C \|\mu_h\| \quad \forall \mu_h \in Q_h.$$

The continuous inf-sup condition was shown in Emerson (2015, Appendix B) and Hu *et al.* (2009, Theorem 3.1). However, the discrete inf-sup condition is not inherited from the continuous problem. Some previous works have succeeded in obtaining a discrete inf-sup condition for some specific discretisations. A discrete inf-sup condition was proven for the $([\mathbb{Q}_k]^d \oplus \mathbb{B}_F)$-$\mathbb{Q}_0$ element on quadrilaterals in Emerson (2015, Lemma 2.5.14) and Adler *et al.* (2015b, Lemma 3.12). The discrete inf-sup condition for the $[\mathbb{P}_1]^2$-$\mathbb{P}_1$ discretisation is shown in Hu *et al.* (2009, Theorem 4.5), where the analysis is only valid for the two-dimensional case due to the use of some special inverse inequalities. It is straightforward to deduce that an enrichment of V_h still guarantees the stability of the discretisation, and thus $[\mathbb{P}_2]^2$-$\mathbb{P}_1$ is inf-sup stable under the same conditions. In three dimensions, there is not yet a discussion about the inf-sup stability of the finite element pair $[\mathbb{P}_2]^3$-$\mathbb{P}_1$ for the bilinear form $\mathfrak{b}(\cdot, \cdot)$, however, we can observe that it is inf-sup stable at least in our numerical experiments in Chapter 4.

We now consider the matrix form of the saddle-point system (2.12) after discretisation. The coercivity of the bilinear form $\mathfrak{a}(\cdot, \cdot)$ implies the invertibility of the coefficient matrix $\mathbf{A}$ and the discrete inf-sup condition indicates that $\mathbf{B}$ has full row rank. We use the full block factorisation preconditioner

$$\mathcal{Q}^{-1} = \begin{bmatrix} \mathbf{I} & -\tilde{\mathbf{A}}^{-1}\mathbf{B}^\top \\ \mathbf{0} & \mathbf{I} \end{bmatrix} \begin{bmatrix} \tilde{\mathbf{A}}^{-1} & \mathbf{0} \\ \mathbf{0} & \tilde{\mathbf{S}}^{-1} \end{bmatrix} \begin{bmatrix} \mathbf{I} & \mathbf{0} \\ -\mathbf{B}\tilde{\mathbf{A}}^{-1} & \mathbf{I} \end{bmatrix},$$

with approximate inner solves $\tilde{\mathbf{A}}^{-1}$ and $\tilde{\mathbf{S}}^{-1}$ for the director block and the Schur complement $\mathbf{S} = -\mathbf{B}\mathbf{A}^{-1}\mathbf{B}^{\top}$, respectively, for solving the saddle-point problem (2.12). With exact inner solves, this is an exact inverse. With this strategy, solving the original saddle-point problem (2.12) reduces to solving two smaller linear systems involving $\mathbf{A}$ and $\mathbf{S}$. Even though $\mathbf{A}$ is sparse, its inverse is generally dense, making it impractical to store $\mathbf{S}$ explicitly. In this situation, developing a fast solver for $\mathbf{A}$ is tractable while approximating $\mathbf{S}$ becomes difficult. We will return to this issue in Section 2.3.2 and Chapter 3.

2.3 Augmented Lagrangian Form

In the previous section, we have considered the method of Lagrange multipliers to enforce the unit-length constraint. We now introduce one of the most famous and successful algorithms, as described in many text books, e.g., Fortin and Glowinski (1983) and Nocedal and Wright (1999), for solving constrained optimisation problems: the augmented Lagrangian method. It can be interpreted as the combination of the pure penalty method and the method of Lagrange multipliers. The AL procedure is to transform the constrained minimisation problem into an unconstrained one by introducing a Lagrange multiplier $\lambda \in L^2(\Omega)$ and adding a term (to its Lagrangian) that penalises the constraint. Instead of solving the constrained problem, we seek the equilibrium of the unconstrained minimisation problem. In this section, we utilise the AL stabilisation strategy and accordingly modify the discrete Newton-linearised system to control the Schur complement.

2.3.1 *Penalising the constraint*

Consider penalising the continuous form of the nonlinear constraint $\mathbf{n} \cdot \mathbf{n} = 1$ in the AL algorithm, then we obtain the associated Lagrangian

$$\tilde{\mathcal{L}}(\mathbf{n}, \lambda) = \mathcal{L}(\mathbf{n}, \lambda) + \frac{\gamma}{2}\left(\mathbf{n} \cdot \mathbf{n} - 1, \mathbf{n} \cdot \mathbf{n} - 1\right)_0, \qquad (2.15)$$

with penalty parameter $\gamma \geq 0$. The weak form of the associated first-order optimality conditions is to find $(\mathbf{n}, \lambda) \in \mathbf{H}_b^1(\Omega) \times L^2(\Omega)$ such that

$$\tilde{\mathcal{L}}_{\mathbf{n}}[\mathbf{v}] = \mathcal{L}_{\mathbf{n}}[\mathbf{v}] + 2\gamma\left(\mathbf{n} \cdot \mathbf{n} - 1, \mathbf{n} \cdot \mathbf{v}\right)_0 = 0 \quad \forall \mathbf{v} \in \mathbf{H}_0^1(\Omega),$$
$$\tilde{\mathcal{L}}_{\lambda}[\mu] = \mathcal{L}_{\lambda}[\mu] = \left(\mu, \mathbf{n} \cdot \mathbf{n} - 1\right)_0 = 0 \qquad \forall \mu \in L^2(\Omega).$$

The Newton linearisation at a given approximation $(\mathbf{n}_j, \lambda_j)$ yields a system of the form:

$$\begin{bmatrix} \tilde{\mathcal{L}}_{\mathbf{n}\mathbf{n}} & \mathcal{L}_{\mathbf{n}\lambda} \\ \mathcal{L}_{\lambda\mathbf{n}} & 0 \end{bmatrix} \begin{bmatrix} \delta\mathbf{n} \\ \delta\lambda \end{bmatrix} = -\begin{bmatrix} \tilde{\mathcal{L}}_{\mathbf{n}} \\ \mathcal{L}_{\lambda} \end{bmatrix}.$$

Thus, we have to solve the augmented discrete variational problem:

$$\begin{aligned} \mathfrak{a}^c(\delta\mathbf{n}_h, \mathbf{v}_h) + \mathfrak{b}(\mathbf{v}_h, \delta\lambda_h) &= \mathfrak{f}^c(\mathbf{v}_h) \quad \forall \mathbf{v}_h \in V_h, \\ \mathfrak{b}(\delta\mathbf{n}_h, \mu_h) &= \mathfrak{g}(\mu_h) \quad \forall \mu_h \in Q_h, \end{aligned} \tag{2.16}$$

where

$$\mathfrak{a}^c(\mathbf{u}, \mathbf{v}) = \mathfrak{a}(\mathbf{u}, \mathbf{v}) + 4\gamma \left(\mathbf{n}_j \cdot \mathbf{u}, \mathbf{n}_j \cdot \mathbf{v}\right)_0 + 2\gamma \left(\mathbf{n}_j \cdot \mathbf{n}_j - 1, \mathbf{u} \cdot \mathbf{v}\right)_0,$$

and

$$\mathfrak{f}^c(\mathbf{v}) = \mathfrak{f}(\mathbf{v}) - 2\gamma \left(\mathbf{n}_j \cdot \mathbf{n}_j - 1, \mathbf{n}_j \cdot \mathbf{v}\right)_0.$$

Comparing (2.16) to the original system (2.13), only the bilinear form $\mathfrak{a}(\cdot, \cdot)$ and the right-hand side functional $\mathfrak{f}(\cdot)$ have changed. The boundedness of $\mathfrak{f}^c(\cdot)$ follows straightforwardly via the Cauchy–Schwarz inequality. As for the coercivity of $\mathfrak{a}^c(\cdot, \cdot)$, an additional assumption on the penalty parameter γ is needed.

Lemma 2.5 (Continuous coercivity). *Let $\alpha_0 > 0$ be the coercivity constant of $\mathfrak{a}(\cdot, \cdot)$. If $\alpha_0 > 2\gamma|\alpha - 1|$ with $0 < \alpha \le 1 \le \beta$ satisfying $\alpha \le |\mathbf{n}_j|^2 \le \beta$, there exists a $\beta_0 > 0$ such that*

$$\mathfrak{a}^c(\mathbf{v}, \mathbf{v}) \ge \beta_0 \|\mathbf{v}\|_1^2 \quad \forall \mathbf{v} \in \mathbf{H}_0^1(\Omega).$$

Proof. Note that

$$\begin{aligned} \mathfrak{a}^c(\mathbf{v}, \mathbf{v}) &= \mathfrak{a}(\mathbf{v}, \mathbf{v}) + 4\gamma \|\mathbf{n}_j \cdot \mathbf{v}\|_0^2 + 2\gamma \left(\mathbf{n}_j \cdot \mathbf{n}_j - 1, \mathbf{v} \cdot \mathbf{v}\right)_0 \\ &\ge \mathfrak{a}(\mathbf{v}, \mathbf{v}) + 2\gamma \left(\mathbf{n}_j \cdot \mathbf{n}_j - 1, \mathbf{v} \cdot \mathbf{v}\right)_0. \end{aligned}$$

By the assumption that $\mathfrak{a}(\mathbf{v}, \mathbf{v}) \ge \alpha_0 \|\mathbf{v}\|_1^2$ for some $\alpha_0 > 0$, we have

$$\mathfrak{a}^c(\mathbf{v}, \mathbf{v}) \ge \alpha_0 \|\mathbf{v}\|_1^2 + 2\gamma \left(\mathbf{n}_j \cdot \mathbf{n}_j - 1, \mathbf{v} \cdot \mathbf{v}\right)_0.$$

Moreover, since $\mathbf{n}_j \cdot \mathbf{n}_j \ge \alpha$ and $\alpha - 1 \le 0$, we get

$$2\gamma \left(\mathbf{n}_j \cdot \mathbf{n}_j - 1, \mathbf{v} \cdot \mathbf{v}\right)_0 \ge 2\gamma(\alpha - 1)\|\mathbf{v}\|_0^2 \ge 2\gamma(\alpha - 1)\|\mathbf{v}\|_1^2.$$

Thus, by taking $\beta_0 = \alpha_0 - 2\gamma|\alpha - 1| > 0$, we obtain the desired coercivity property. $\qquad \square$

The condition $\alpha_0 > 2\gamma|\alpha - 1|$ in Lemma 2.5 indicates a limit on the value of γ to ensure the solvability of the augmented system (2.16). However, it is desirable to use large values of γ to achieve better control of the Schur complement as we shall see in Chapter 4. We therefore choose to employ a Picard iteration to solve the nonlinear problem, omitting the term $2\gamma\left(\mathbf{n}_j \cdot \mathbf{n}_j - 1, \mathbf{v} \cdot \mathbf{v}\right)_0$ from the linearised equations. This yields the linearised problem: find $(\delta\mathbf{n}_h, \delta\lambda_h) \in V_h \times Q_h$ such that

$$
\begin{aligned}
\mathfrak{a}^m(\delta\mathbf{n}_h, \mathbf{v}_h) + \mathfrak{b}(\mathbf{v}_h, \delta\lambda_h) &= \mathfrak{f}^c(\mathbf{v}_h) \quad \forall \mathbf{v}_h \in V_h, \\
\mathfrak{b}(\delta\mathbf{n}_h, \mu_h) &= \mathfrak{g}(\mu_h) \quad \forall \mu_h \in Q_h,
\end{aligned}
\tag{2.17}
$$

with the modified bilinear form

$$
\mathfrak{a}^m(\mathbf{u}, \mathbf{v}) = \mathfrak{a}(\mathbf{u}, \mathbf{v}) + 4\gamma\left(\mathbf{n}_j \cdot \mathbf{u}, \mathbf{n}_j \cdot \mathbf{v}\right)_0
\tag{2.18}
$$

to be solved at each nonlinear iteration. This ensures that the $(1,1)$-block is coercive with a coercivity constant independent of γ. Moreover, in contrast to the situation with the Navier–Stokes equations, numerical experiments indicate that the use of this Picard iteration requires *fewer* nonlinear iterations to converge to a given tolerance than using the full Newton linearisation (see Section 4.2.1).

The corresponding matrix form of the variational problem (2.17) becomes

$$
\begin{bmatrix} \mathbf{A} + \gamma\mathbf{A}_* & \mathbf{B}^\top \\ \mathbf{B} & \mathbf{0} \end{bmatrix} \begin{bmatrix} U \\ X \end{bmatrix} = \begin{bmatrix} \mathbf{f} + \gamma\mathbf{l} \\ \mathbf{g} \end{bmatrix},
\tag{2.19}
$$

where $\mathbf{A}_*$ is the assembly of $4\left(\mathbf{n}_j \cdot \mathbf{u}, \mathbf{n}_j \cdot \mathbf{v}\right)_0$ and $\mathbf{l}$ denotes the assembly of $-2\left(\mathbf{n}_j \cdot \mathbf{n}_j - 1, \mathbf{n}_j \cdot \mathbf{v}\right)_0$. Note that compared to the non-augmented version (2.12), the $(1,1)$ block in (2.19) has an additional semi-definite term $\gamma\mathbf{A}_*$ with a large coefficient γ. Its sparsity pattern remains unchanged. We will construct a robust multigrid method to solve this top-left block in Chapter 3.

Since the unit-length constraint is enforced exactly in (2.15), the continuous solutions to minimising both (2.15) and (2.8) are the same. However, the unit-length constraint is not enforced exactly in our finite element discretisation, and hence this AL stabilisation does change the computed discrete solution.

Remark 2.7. When utilising the augmented Lagrangian strategy, one can apply it before discretisation or afterwards. In this part of work we consider the continuous penalisation, as it improves the enforcement of the nonlinear constraint, as shown later in Section 2.3.3. This is different to the approach considered in Benzi and Olshanskii (2006) and Farrell *et al.* (2019) for the stationary Navier–Stokes equations, where the discrete AL stabilisation was used to yield a system that has the same discrete solution but a better Schur complement.

2.3.2 *Approximation to the Schur complement*

The Schur complement of the augmented director block in (2.19) is given by

$$\mathbf{S}_\gamma = -\mathbf{B}\mathbf{A}_\gamma^{-1}\mathbf{B}^\top =: -\mathbf{B}(\mathbf{A} + \gamma\mathbf{A}_*)^{-1}\mathbf{B}^\top.$$

We now proceed to analyse this Schur complement by following similar techniques to those of Heister and Rapin (2012, §4). We will show that $\mathbf{A}_*$ is equal to the matrix arising from the *discrete* AL stabilisation (which controls the Schur complement) plus a perturbation that vanishes as the mesh is refined.

Let $\Pi_{Q_h} : L^2(\Omega) \to Q_h$ (Q_h is a finite-dimensional approximation space of $L^2(\Omega)$) be the orthogonal L^2 projection operator, i.e., there holds for $p \in L^2(Omega)$ that

$$(p - \Pi_{Q_h}p, q)_0 = 0 \quad \forall q \in Q_h.$$

We define the fluctuation operator $\mathfrak{F} := \mathcal{I} - \Pi_{Q_h}$ where $\mathcal{I} : L^2(\Omega) \to L^2(\Omega)$ is the identity mapping. Therefore, one has

$$(\mathfrak{F}(p), q)_0 = 0 \quad \forall q \in Q_h.$$

For $\mathbf{u}_h, \mathbf{v}_h \in V_h$, one can split the term $4\,(\mathbf{n}_j \cdot \mathbf{u}_h, \mathbf{n}_j \cdot \mathbf{v})_0$ into the following terms using the properties of $\mathfrak{F}$ and Π_{Q_h}:

$$
\begin{aligned}
4\,(\mathbf{n}_j \cdot \mathbf{u}, \mathbf{n}_j \cdot \mathbf{v})_0 &= (\Pi_{Q_h}(2\mathbf{n}_j \cdot \mathbf{n}), 2\mathbf{n}_j \cdot \mathbf{v})_0 + (\mathfrak{F}(2\mathbf{n}_j \cdot \mathbf{u}), 2\mathbf{n}_j \cdot \mathbf{v})_0 \\
&= (\Pi_{Q_h}(2\mathbf{n}_j \cdot \mathbf{n}), (\Pi_{Q_h} + \mathfrak{F})(2\mathbf{n}_j \cdot \mathbf{v}))_0 \\
&\quad + (\mathfrak{F}(2\mathbf{n}_j \cdot \mathbf{u}), (\Pi_{Q_h} + \mathfrak{F})(2\mathbf{n}_j \cdot \mathbf{v}))_0 \\
&= (\Pi_{Q_h}(2\mathbf{n}_j \cdot \mathbf{u}), \Pi_{Q_h}(2\mathbf{n}_j \cdot \mathbf{v}))_0 + (\mathfrak{F}(2\mathbf{n}_j \cdot \mathbf{u}), \mathfrak{F}(2\mathbf{n}_j \cdot \mathbf{v}))_0.
\end{aligned}
$$

Note here that the assembly of the first term is $\mathbf{B}^\top\mathbf{M}_\lambda^{-1}\mathbf{B}$, where $\mathbf{M}_\lambda$ is the mass matrix associated with the finite element space Q_h for the multiplier. This can then be readily used with the Sherman–Morrison–Woodbury formula to derive an approximation of the Schur complement. Moreover, the second term $(\mathfrak{F}(2\mathbf{n}_j \cdot \mathbf{u}), \mathfrak{F}(2\mathbf{n}_j \cdot \mathbf{v}))_0$ in fact characterises the difference between $\mathbf{A}_*$ and $\mathbf{B}^\top \mathbf{M}_\lambda^{-1} \mathbf{B}$, since the assembly of $4(\mathbf{n}_j \cdot \mathbf{u}, \mathbf{n}_j \cdot \mathbf{v})_0$ is $\mathbf{A}_*$. The next result (see Theorem 2.6) shows that such difference vanishes as the mesh size $h \to 0$ and thus, in this limit, the tractable term $\mathbf{B}^\top \mathbf{M}_\lambda^{-1} \mathbf{B}$ dominates $\mathbf{A}_*$.

Theorem 2.6. *Let $(\delta \mathbf{n}_h, \delta \lambda_h) \in V_h \times Q_h$ be the solution of the augmented discrete system (2.17) with corresponding degrees of freedom $(U, X) \in \mathbb{R}^n \times \mathbb{R}^m$. Assume that $\|\delta \mathbf{n}_h\|_1$ is bounded as $h \to 0$. Then, for the Newton linearisation at a given approximation $(\mathbf{n}_j, \lambda_j)$ satisfying $\alpha \le |\mathbf{n}_j|^2 \le \beta$ with $0 < \alpha \le 1 \le \beta$ and $|\nabla \mathbf{n}_j|$ bounded pointwise a.e., we have*

$$\left\| \left(\mathbf{A}_* - \mathbf{B}^\top \mathbf{M}_\lambda^{-1} \mathbf{B} \right) U \right\|_{\mathbb{R}^n} \lesssim h^{1 + \frac{d}{2}} \|\delta \mathbf{n}_h\|_1,$$

where $\| \cdot \|_{\mathbb{R}^n}$ denotes the Euclidean norm.

Proof. Assuming $\mathbf{v}_h \in V_h$ and using the basis representations in $V_h = \text{span}\{\varphi_i\}$ for $\delta \mathbf{n}_h$ and $\mathbf{v}_h$:

$$\delta \mathbf{n}_h = \sum_{i=1}^n U_i \varphi_i, \quad \mathbf{v}_h = \sum_{i=1}^n Y_i \varphi_i,$$

we obtain

$$\left\| \left(\mathbf{A}_* - \mathbf{B}^\top \mathbf{M}_\lambda^{-1} \mathbf{B} \right) U \right\|_{\mathbb{R}^n} = \sup_{\|Y\|_{\mathbb{R}^n} = 1} Y^\top \left(\mathbf{A}_* - \mathbf{B}^\top \mathbf{M}_\lambda^{-1} \mathbf{B} \right) U$$

$$= \sup_{\substack{\mathbf{v}_h = \sum_{i=1}^n Y_i \varphi_i \\ \|Y\|_{\mathbb{R}^n} = 1}} (\mathfrak{F}(2\mathbf{n}_j \cdot \delta \mathbf{n}_h), \mathfrak{F}(2\mathbf{n}_j \cdot \mathbf{v}_h))_0$$

$$\le \sup_{\substack{\mathbf{v}_h = \sum_{i=1}^n Y_i \varphi_i \\ \|Y\|_{\mathbb{R}^n} = 1}} \|\mathfrak{F}(2\mathbf{n}_j \cdot \delta \mathbf{n}_h)\|_0 \|\mathfrak{F}(2\mathbf{n}_j \cdot \mathbf{v}_h)\|_0$$

$$\le \underbrace{\|\mathfrak{F}\|}_{G_1} \underbrace{\sup_{\substack{\mathbf{v}_h = \sum_{i=1}^n Y_i \varphi_i \\ \|Y\|_{\mathbb{R}^n} = 1}} \|2\mathbf{n}_j \cdot \mathbf{v}_h\|_0}_{G_2} \underbrace{\|\mathfrak{F}(2\mathbf{n}_j \cdot \delta \mathbf{n}_h)\|_0}_{G_3}$$

by applying the Cauchy–Schwarz inequality.

One readily sees that $G_1 \le C_1$ for a certain constant C_1 from the continuity of $\mathfrak{F}$. Furthermore, we write

$$G_2 = \sup_{\mathbf{v}_h = \sum_{i=1}^n Y_i \varphi_i} \frac{\|2\mathbf{n}_j \cdot \mathbf{v}_h\|_0}{\|Y\|_{\mathbb{R}^n}}.$$

Note that (Knabner and Angermann, 2000, Theorem 3.43) as used in Heister and Rapin (2012) gives the relation between the discrete vector Y and its associated continuous function $\mathbf{v}_h$:

$$\|Y\|_{\mathbb{R}^n} \geq C_r h^{-\frac{d}{2}} \|\mathbf{v}_h\|_0,$$

for some $C_r > 0$. Then with the fact that $\mathbf{n}_j$ is bounded we have

$$G_2 \leq \sup_{\mathbf{v}_h} \frac{\|2\mathbf{n}_j \cdot \mathbf{v}_h\|_0}{C_r h^{-\frac{d}{2}} \|\mathbf{v}_h\|_0} \leq C_2 h^{\frac{d}{2}}.$$

Moreover, Clément (1975, Theorem 1) implies

$$\|\mathfrak{F}(p)\|_0 = \|p - \Pi_{Q_h} p\|_0 \leq C_4 h \|p\|_1 \quad \forall p \in H^1(\Omega),$$

and we can deduce the following L^2-projection error estimate

$$G_3 = \|\mathfrak{F}(2\mathbf{n}_j \cdot \delta\mathbf{n}_h)\|_0 \leq C_4 h \|2\mathbf{n}_j \cdot \delta\mathbf{n}_h\|_1 \leq C_3 h \|\delta\mathbf{n}_h\|_1.$$

Note here we have used the pointwise boundedness of $\mathbf{n}_j, \nabla\mathbf{n}_j$ a.e. and the fact that $\delta\mathbf{n}_h \in V_h \subset H^1(\Omega)$.

Combining these estimates regarding G_1, G_2, G_3, we find

$$\left\|\left(\mathbf{A}_* - \mathbf{B}^\top \mathbf{M}_\lambda^{-1} \mathbf{B}\right) U\right\|_{\mathbb{R}^n} \lesssim h^{1+\frac{d}{2}} \|\delta\mathbf{n}_h\|_1 \to 0 \quad \text{as } h \to 0.$$

The proof is complete. $\qquad\qquad\qquad\qquad\qquad\qquad\qquad\qquad\square$

This result suggests the use of the algebraic approximation

$$\mathbf{S}_\gamma \approx -\mathbf{B}\left(\mathbf{A} + \gamma\mathbf{B}^\top \mathbf{M}_\lambda^{-1} \mathbf{B}\right)^{-1} \mathbf{B}^\top. \tag{2.20}$$

The reason for doing so is that we can straightforwardly calculate the inverse (note the solver requires the action of $\mathbf{S}_\gamma^{-1}$, i.e., solving linear systems involving $\mathbf{S}_\gamma$) of this approximation (2.20) by the Sherman–Morrison–Woodbury formula as shown in the following Lemma 2.7.

Lemma 2.7. *The Schur complement approximation satisfies*

$$\mathbf{S}_\gamma^{-1} = \mathbf{S}^{-1} - \gamma\mathbf{M}_\lambda^{-1}. \tag{2.21}$$

Proof. Recalling the Sherman–Morrison–Woodbury formula (Hager, 1989): for matrices $\mathbf{E}, \mathbf{U}_1, \mathbf{P}$ and $\mathbf{U}_2$ where $\mathbf{E}, \mathbf{P}$ and $\mathbf{P}^{-1} + \mathbf{U}_2\mathbf{E}^{-1}\mathbf{U}_1$

are invertible, it holds that

$$(\mathbf{E} + \mathbf{U}_1 \mathbf{P} \mathbf{U}_2)^{-1} = \mathbf{E}^{-1} - \mathbf{E}^{-1} \mathbf{U}_1 \left(\mathbf{P}^{-1} + \mathbf{U}_2 \mathbf{E}^{-1} \mathbf{U}_1 \right)^{-1} \mathbf{U}_2 \mathbf{E}^{-1}. \quad (2.22)$$

We now apply this formula twice to obtain

$$\mathbf{S}_\gamma^{-1} = \left(-\mathbf{B} \left(\mathbf{A} + \gamma \mathbf{B}^\top \mathbf{M}_\lambda^{-1} \mathbf{B} \right)^{-1} \mathbf{B}^\top \right)^{-1}$$

$$\overset{(2.22)}{=} - \left(\mathbf{B} \left(\mathbf{A}^{-1} - \mathbf{A}^{-1} \mathbf{B}^\top \left(\frac{1}{\gamma} \mathbf{M}_\lambda + \mathbf{B} \mathbf{A}^{-1} \mathbf{B}^\top \right)^{-1} \mathbf{B} \mathbf{A}^{-1} \right) \mathbf{B}^\top \right)^{-1}$$

$$= - \left(\underbrace{\mathbf{B} \mathbf{A}^{-1} \mathbf{B}^\top}_{-\mathbf{S}} - \underbrace{\mathbf{B} \mathbf{A}^{-1} \mathbf{B}^\top}_{-\mathbf{S}} \left(\frac{1}{\gamma} \mathbf{M}_\lambda + \underbrace{\mathbf{B} \mathbf{A}^{-1} \mathbf{B}^\top}_{-\mathbf{S}} \right)^{-1} \underbrace{\mathbf{B} \mathbf{A}^{-1} \mathbf{B}^\top}_{-\mathbf{S}} \right)^{-1}$$

$$= \left(\mathbf{S} + \mathbf{S} \left(\frac{1}{\gamma} \mathbf{M}_\lambda - \mathbf{S} \right)^{-1} \mathbf{S} \right)^{-1}$$

$$\overset{(2.22)}{=} \mathbf{S}^{-1} - \mathbf{S}^{-1} \mathbf{S} \left(\frac{1}{\gamma} \mathbf{M}_\lambda - \mathbf{S} + \mathbf{S} \mathbf{S}^{-1} \mathbf{S} \right)^{-1} \mathbf{S} \mathbf{S}^{-1}$$

$$= \mathbf{S}^{-1} - \gamma \mathbf{M}_\lambda^{-1}.$$

This completes the proof.   □

Induced from the above result (2.21) for the inverse of the Schur complement approximation, a simple and effective approach for large γ is to employ the approximation

$$\mathbf{S}_\gamma^{-1} \approx -\gamma \mathbf{M}_\lambda^{-1}. \quad (2.23)$$

On the infinite-dimensional level, the effect of the augmented Lagrangian term is to make $-\gamma^{-1} \mathcal{I}$ ($\mathcal{I}$ the identity operator on the multiplier space) an effective approximation for the Schur complement (Polyak and Tret'yakov, 1974, Lemma 3). When discretised, this indicates that the weighted multiplier mass matrix $-\gamma^{-1} \mathbf{M}_\lambda$ will be an effective approximation for $\mathbf{S}_\gamma$, with the approximation improving as $\gamma \to \infty$.

In fact, the approximation of the inverse of the discretely augmented Schur complement (2.23) can be improved further by combining $-\gamma \mathbf{M}_\lambda^{-1}$ with a good approximation of the unaugmented Schur complement $\mathbf{S}$

(He *et al.*, 2018). Given an approximation $\tilde{\mathbf{S}}$ of $\mathbf{S}$, we employ

$$\mathbf{S}_\gamma^{-1} \approx \tilde{\mathbf{S}}_\gamma^{-1} = \tilde{\mathbf{S}}^{-1} - \gamma \mathbf{M}_\lambda^{-1}. \tag{2.24}$$

It is therefore of interest to consider the Schur complement of the unaugmented problem. In the context of the Stokes equations, the Schur complement is spectrally equivalent to the viscosity-weighted pressure mass matrix (Elman *et al.*, 2014; Silvester and Wathen, 1994; Wathen and Silvester, 1991). Following similar techniques, an approximation can be obtained by proving that $\mathbf{B}\mathbf{A}^{-1}\mathbf{B}^\top$ is spectrally equivalent to $\mathbf{M}_\lambda$ for the equal-constant nematic case. This gives us good insight into the choice of $\tilde{\mathbf{S}}^{-1}$.

Theorem 2.8. *Assume that the finite-dimensional spaces $V_h \subset \mathbf{H}_0^1(\Omega)$ and $Q_h \subset L^2(\Omega)$ are inf-sup stable. For equal-constant nematic LC problems without augmented Lagrangian stabilisation, the matrix $\mathbf{B}\mathbf{A}^{-1}\mathbf{B}^\top$ arising from the Newton-linearised system is spectrally equivalent to the multiplier mass matrix $\mathbf{M}_\lambda$, under the same assumptions as in Lemma 2.4.*

Proof. For the equal-constant model with Dirichlet boundary conditions $\mathbf{n} = \mathbf{n}_b \in H^{1/2}(\partial\Omega, \mathcal{S}^2)$, its corresponding Lagrangian is

$$\mathcal{L}(\mathbf{n}, \lambda) = \frac{K_c}{2}\left(\nabla\mathbf{n}, \nabla\mathbf{n}\right)_0 + \left(\lambda, \mathbf{n}\cdot\mathbf{n} - 1\right)_0.$$

After Newton linearisation and due to the inf-sup stability of the finite element pair V_h-Q_h, the discrete variational problem is to find $\delta\mathbf{n}_h \in V_h$, $\delta\lambda_h \in Q_h$ satisfying

$$K_c\left(\nabla\delta\mathbf{n}_h, \nabla\mathbf{v}_h\right)_0 + 2\left(\lambda_j, \delta\mathbf{n}_h \cdot \mathbf{v}_h\right)_0 + 2\left(\delta\lambda_h, \mathbf{n}_j \cdot \mathbf{v}_h\right)_0$$
$$= -K_c\left(\nabla\mathbf{n}_j \cdot \nabla\mathbf{v}_h\right)_0 - 2\left(\lambda_j, \mathbf{n}_j \cdot \mathbf{v}_h\right)_0 \quad \forall \mathbf{v}_h \in V_h,$$
$$2\left(\mu_h, \mathbf{n}_j \cdot \delta\mathbf{n}_h\right)_0 = -\left(\mu_h, \mathbf{n}_j \cdot \mathbf{n}_j - 1\right)_0 \quad \forall \mu_h \in Q_h,$$

where $\mathbf{n}_j$ and λ_j represent the current approximations to $\mathbf{n}$ and λ, respectively. This can be rewritten in block matrix form as

$$\mathcal{R}\begin{bmatrix} U \\ X \end{bmatrix} := \begin{bmatrix} \mathbf{A} & \mathbf{B}^\top \\ \mathbf{B} & 0 \end{bmatrix}\begin{bmatrix} U \\ X \end{bmatrix} = \begin{bmatrix} \mathbf{f} \\ \mathbf{g} \end{bmatrix},$$

where as before $U \in \mathbb{R}^n$ and $X \in \mathbb{R}^m$ are the unknown coefficients of the discrete director update and the discrete Lagrange multiplier update with respect to the basis functions in V_h and Q_h, and $\mathbf{A}$ denotes the symmetric form $K_c\left(\nabla\delta\mathbf{n}_h, \nabla\mathbf{v}_h\right)_0 + 2\left(\lambda_j, \delta\mathbf{n}_h \cdot \mathbf{v}_h\right)_0$. The coercivity property of the bilinear form from Lemma 2.4 ensures that $\mathbf{A}$ is positive definite.

The coefficient matrix $\mathcal{R}$ is symmetric and indefinite (resulting in $\mathcal{R}$ possessing both positive and negative eigenvalues). Moreover, $\mathcal{R}$ is non-singular if and only if $\mathbf{B}$ has full row rank, which can be deduced from the discrete inf-sup condition.

Denote

$$\|\mathbf{u}_h\|_{lc}^2 = K_c (\nabla \mathbf{u}_h, \nabla \mathbf{u}_h)_0 + (\lambda_j, 2\mathbf{u}_h \cdot \mathbf{u}_h)_0,$$

$$\|\mu_h\|_0^2 = (\mu_h, \mu_h)_0.$$

Notice that the validity of the first norm follows from the assumed pointwise non-negativity of λ_j.

For a stable mixed finite element, from the inf-sup condition, there exists a positive constant C independent of the mesh size h such that

$$\sup_{\mathbf{u}_h \in V_h \backslash \{0\}} \frac{(\mu_h, 2\mathbf{n}_j \cdot \mathbf{u}_h)_0}{\|\mathbf{u}_h\|_{lc}} \geq C \|\mu_h\|_0 \quad \forall \mu_h \in Q_h,$$

leading to its matrix form

$$\max_{U \in \mathbb{R}^n \backslash \{0\}} \frac{X^\top \mathbf{B} U}{[U^\top \mathbf{A} U]^{1/2}} \geq C [X^\top \mathbf{M}_\lambda X]^{1/2} \quad \forall X \in \mathbb{R}^m.$$

Thus, we have

$$C [X^\top \mathbf{M}_\lambda X]^{1/2} \leq \max_{U \in \mathbb{R}^n \backslash \{0\}} \frac{X^\top \mathbf{B} U}{[U^\top \mathbf{A} U]^{1/2}}$$

$$= \max_{\mathbf{z} = A^{1/2} U \neq 0} \frac{X^\top \mathbf{B} \mathbf{A}^{-1/2} \mathbf{z}}{[\mathbf{z}^\top \mathbf{z}]^{1/2}}$$

$$= (X^\top \mathbf{B} \mathbf{A}^{-1} \mathbf{B}^\top X)^{1/2} \quad \forall X \in \mathbb{R}^m,$$

where the maximum is attained at $\mathbf{z} = (X^\top \mathbf{B} \mathbf{A}^{-1/2})^\top$. It yields

$$C^2 \frac{X^\top \mathbf{M}_\lambda X}{X^\top X} \leq \frac{X^\top \mathbf{B} \mathbf{A}^{-1} \mathbf{B}^\top X}{X^\top X} \quad \forall X \in \mathbb{R}^m \backslash \{0\}. \tag{2.25}$$

Regardless of the stability of the finite element pair, we can deduce from the boundedness of $\mathfrak{b}(\cdot, \cdot)$ that there exists a positive constant C_1 such that

$$X^\top \mathbf{B} U \leq C_1 [X^\top \mathbf{M}_\lambda X]^{1/2} [U^\top \mathbf{A} U]^{1/2} \quad \forall U \in \mathbb{R}^n, \forall X \in \mathbb{R}^m.$$

Hence,

$$C_1[X^\top \mathbf{M}_\lambda X]^{1/2} \geq \max_{U \in \mathbb{R}^n \setminus \{0\}} \frac{X^\top \mathbf{B} U}{[U^\top \mathbf{A} U]^{1/2}}$$

$$= \max_{\mathbf{z} = \mathbf{A}^{1/2} U \neq 0} \frac{X^\top \mathbf{B} \mathbf{A}^{-1/2} \mathbf{z}}{[\mathbf{z}^\top \mathbf{z}]^{1/2}}$$

$$= (X^\top \mathbf{B} \mathbf{A}^{-1} \mathbf{B}^\top X)^{1/2} \quad \forall X \in \mathbb{R}^m,$$

where again the maximum is attained at $\mathbf{z} = (X^\top \mathbf{B} \mathbf{A}^{-1/2})^\top$. This gives rise to

$$\frac{X^\top \mathbf{B} \mathbf{A}^{-1} \mathbf{B}^\top X}{X^\top \mathbf{M}_\lambda X} \leq C_1^2 \quad \forall X \in \mathbb{R}^m \setminus \{0\}. \tag{2.26}$$

Therefore for inf-sup stable finite element pairs, we have by (2.25) and (2.26)

$$C^2 \leq \frac{X^\top \mathbf{B} \mathbf{A}^{-1} \mathbf{B}^\top X}{X^\top \mathbf{M}_\lambda X} \leq C_1^2 \quad \forall X \in \mathbb{R}^m \setminus \{0\}.$$

This indicates that $\mathbf{B} \mathbf{A}^{-1} \mathbf{B}^\top$ is spectrally equivalent to $\mathbf{M}_\lambda$. □

Remark 2.8. It follows from Theorem 2.8 that $\gamma = 0$ should show mesh-independence (i.e., the average number of Flexible GMRES (FGMRES; this allows for the use of different preconditioner in each iteration step) iterations per nonlinear iteration does not deteriorate as one refines the mesh) in the case of equal-constant nematic LC. This can be observed in subsequent numerical experiments reported in Table 4.6 (see the column where $\gamma = 0$). One should also notice that such mesh-independence for $\gamma = 0$ is also shown in Table 4.2 for the non-equal-constant case, suggesting it has use outside the context of augmented Lagrangian methods also.

Combining Theorem 2.8 with (2.24), our final approximation for $\mathbf{S}_\gamma^{-1}$ is given by

$$\mathbf{S}_\gamma^{-1} \approx \tilde{\mathbf{S}}_\gamma^{-1} = -(1 + \gamma) \mathbf{M}_\lambda^{-1}. \tag{2.27}$$

2.3.3 *Improvement of the constraint*

We have now observed that the continuous AL form introduced in Section 2.3.2 can help control the Schur complement. Another contribution of this AL stabilisation is that it improves the discrete enforcement of the constraint as we increase the value of the penalty parameter γ. An example of improving the linear divergence-free constraint in the Stokes system can be found in John *et al.* (2017, Section 5.1). In this section, we will use a similar strategy to show the improvement of the discrete constraint as γ increases.

We restrict ourselves to the equal-constant case with *constant* Dirichlet boundary conditions. That is to say, we consider the Oseen–Frank model with Dirichlet boundary condition $\mathbf{n}|_{\partial\Omega} = \mathbf{n}_b$, where $\mathbf{n}_b$ is a nonzero constant vector satisfying $|\mathbf{n}_b| = 1$.

Remark 2.9. One may wonder whether the solution under this assumption of boundary conditions is the constant boundary data itself, i.e., $\mathbf{n}_h = \mathbf{n}_b$ in the domain Ω, and thus the unit-length constraint is actually satisfied exactly. In fact, $\mathbf{n}_h = \mathbf{n}_b$ is indeed an equilibrium of the energy minimisation problem (2.1), however, it is not the only one and not in general the one with the lowest energy value. An example supports this fact can be seen in Emerson *et al.* (2018), where $\mathbf{n} = (0, 0, 1)$ is strongly enforced on the boundary while many computed solutions are not $(0, 0, 1)$ everywhere in the domain Ω.

We use the $[\mathbb{P}_1]^d$-$\mathbb{P}_1$ finite element pair in this section, so both the director $\mathbf{n}$ and the Lagrange multiplier λ are approximated by continuous piecewise-linear polynomials. For this section, we denote finite element spaces for the director and the Lagrange multiplier by $V_{h,b} := V_h \cap \mathbf{H}_b^1(\Omega)$ and $Q_h \subset L^2(\Omega)$, respectively, and denote $V_{h,0} = V_h \cap \mathbf{H}_0^1(\Omega)$.

We restate the associated nonlinear discrete variational problem as follows: find $(\mathbf{n}_h, \lambda_h) \in V_{h,b} \times Q_h$ such that

$$K_c \left(\nabla \mathbf{n}_h, \nabla \mathbf{v}_h \right)_0 + K_c q_0 \left(\mathbf{v}_h, \nabla \times \mathbf{n}_h \right)_0 + K_c q_0 \left(\mathbf{n}_h, \nabla \times \mathbf{v}_h \right)_0$$
$$+ 2 \left(\lambda_h, \mathbf{n}_h \cdot \mathbf{v}_h \right)_0 + 2\gamma \left(\mathbf{n}_h \cdot \mathbf{n}_h - 1, \mathbf{n}_h \cdot \mathbf{v}_h \right)_0 = 0 \quad \forall \mathbf{v}_h \in V_{h,0}, \tag{2.28a}$$

$$\left(\mu_h, \mathbf{n}_h \cdot \mathbf{n}_h - 1 \right)_0 = 0 \quad \forall \mu_h \in Q_h. \tag{2.28b}$$

Take the test function $\mathbf{v}_h = \mathbf{n}_h - \mathbf{n}_b \in V_{h,0}$ in (2.28a) to obtain

$$K_c \|\nabla \mathbf{n}_h\|_0^2 + 2K_c q_0 \left(\mathbf{n}_h, \nabla \times \mathbf{n}_h \right)_0$$
$$+ 2 \left(\lambda_h, \mathbf{n}_h \cdot \mathbf{n}_h \right)_0 + 2\gamma \left(\mathbf{n}_h \cdot \mathbf{n}_h - 1, \mathbf{n}_h \cdot \mathbf{n}_h \right)_0$$
$$= K_c q_0 \left(\mathbf{n}_b, \nabla \times \mathbf{n}_h \right)_0$$
$$+ 2 \left(\lambda_h, \mathbf{n}_h \cdot \mathbf{n}_b \right)_0 + 2\gamma \left(\mathbf{n}_h \cdot \mathbf{n}_h - 1, \mathbf{n}_h \cdot \mathbf{n}_b \right)_0. \tag{2.29}$$

Note that in this step we have used the fact that since $\mathbf{n}_b$ is a constant vector, its derivative is zero.

As (2.28b) is valid for arbitrary $\mu_h \in Q_h$ and one can easily verify that $\mathbf{n}_h \cdot \mathbf{n}_b \in Q_h$, we have

$$(\mathbf{n}_h \cdot \mathbf{n}_b, \mathbf{n}_h \cdot \mathbf{n}_h - 1)_0 = 0.$$

Then taking $\mu_h = 1$ and $\mu_h = \lambda_h$ leads to

$$(1, \mathbf{n}_h \cdot \mathbf{n}_h - 1)_0 = 0 \quad \text{and} \quad (\lambda_h, \mathbf{n}_h \cdot \mathbf{n}_h - 1)_0 = 0,$$

respectively. Thus, (2.29) collapses to

$$K_c \|\nabla \mathbf{n}_h\|_0^2 + 2K_c q_0 (\mathbf{n}_h, \nabla \times \mathbf{n}_h)_0 + 2 (\lambda_h, 1)_0 + 2\gamma \|\mathbf{n}_h \cdot \mathbf{n}_h - 1\|_0^2$$
$$= K_c q_0 (\mathbf{n}_b, \nabla \times \mathbf{n}_h)_0 + 2 (\lambda_h, \mathbf{n}_h \cdot \mathbf{n}_b)_0. \tag{2.30}$$

By the Cauchy–Schwarz and Hölder inequalities, we observe an upper bound for the right-hand side of (2.30):

$$K_c q_0 (\mathbf{n}_b, \nabla \times \mathbf{n}_h)_0 + 2 (\lambda_h, \mathbf{n}_h \cdot \mathbf{n}_b)_0 \leq K_c q_0 \|\nabla \times \mathbf{n}_h\|_0 + 2 \|\lambda_h\|_0 \|\mathbf{n}_h\|_0$$
$$\leq \frac{K_c q_0}{2} + \frac{K_c q_0}{2} \|\nabla \times \mathbf{n}_h\|_0^2$$
$$+ \|\lambda_h\|_0^2 + \|\mathbf{n}_h\|_0^2. \tag{2.31}$$

Meanwhile, the left-hand side of (2.30) can be bounded from below:

$$K_c \|\nabla \mathbf{n}_h\|_0^2 + 2K_c q_0 (\mathbf{n}_h, \nabla \times \mathbf{n}_h)_0 + 2 (\lambda_h, 1)_0 + 2\gamma \|\mathbf{n}_h \cdot \mathbf{n}_h - 1\|_0^2$$
$$\geq K_c \|\nabla \mathbf{n}_h\|_0^2 - 2K_c q_0 | (\mathbf{n}_h, \nabla \times \mathbf{n}_h)_0 |$$
$$- 2| (\lambda_h, 1)_0 | + 2\gamma \|\mathbf{n}_h \cdot \mathbf{n}_h - 1\|_0^2$$
$$\geq K_c \|\nabla \mathbf{n}_h\|_0^2 - K_c q_0 \|\mathbf{n}_h\|_0^2 - K_c q_0 \|\nabla \times \mathbf{n}_h\|_0^2$$
$$- \|\lambda_h\|_0^2 - |\Omega| + 2\gamma \|\mathbf{n}_h \cdot \mathbf{n}_h - 1\|_0^2, \tag{2.32}$$

where $|\Omega|$ denotes the measure of the domain Ω.

Hence, by combining (2.31) and (2.32), we have

$$K_c \|\nabla \mathbf{n}_h\|_0^2 - (K_c q_0 + 1) \|\mathbf{n}_h\|_0^2 - \frac{3}{2} K_c q_0 \|\nabla \times \mathbf{n}_h\|_0^2$$
$$- \|\lambda_h\|_0^2 + 2\gamma \|\mathbf{n}_h \cdot \mathbf{n}_h - 1\|_0^2 \leq \frac{K_c q_0}{2} + |\Omega|. \tag{2.33}$$

Note that the right-hand side of (2.33) is a fixed constant independent of γ and those negative terms on the left-hand side actually depends on γ since both $\mathbf{n}_h$ and λ_h depends on γ. Therefore, taking γ larger value does not directly force the constraint approximation error $\|\mathbf{n}_h \cdot \mathbf{n}_h - 1\|_0$ to become smaller. That is to say, (2.33) does not imply that $\|\mathbf{n}_h \cdot \mathbf{n}_h - 1\|_0 \leq \mathcal{O}(\gamma^{-1/2})$. However, this improvement of the discrete constraint as γ increases can be observed in our numerical experiments illustrated in Chapter 4.

Remark 2.10. The technique shown in this section can be extended in a similar way to the multi-constant case; we omit the details here for brevity.

2.4   Summary

In this chapter, we considered the Oseen–Frank model of cholesteric LC, which demands a unit-length constraint be enforced. We then applied the continuous augmented Lagrangian form for constraint penalisation and illustrated its two major effects: the improvement of the discrete constraint $\mathbf{n}_h \cdot \mathbf{n}_h = 1$, and a better control on the Schur complement using a weighted mass matrix approximation. However, this results in a more complicated top-left block to be solved, which we will tackle by means of a robust multi-grid method in the next chapter.

Chapter 3

A Robust Multigrid Algorithm for the Augmented Director Block

As discussed in the previous chapter, the addition of the augmented Lagrangian term gives a better approximation to the Schur complement (we will see this in Tables 4.2 and 4.6). However, the tradeoff is that it complicates the solution of the top-left block $\mathbf{A}_\gamma$, as it adds a semi-definite term with a large coefficient. We demonstrate this effect in Table 3.1 where we apply the block preconditioner with the Schur complement approximation $\tilde{\mathbf{S}}_\gamma^{-1}$ as given by (2.27) and $\mathbf{A}_\gamma = \mathbf{A} + \gamma \mathbf{A}_*$ solved approximately with one V-cycle of standard geometric multigrid with Jacobi relaxation. Table 3.1 shows that the solver is neither γ- or h-robust. Thus, for the augmented Lagrangian strategy to be successful, we require a parameter-independent solver for the top-left block.

Fortunately, a rich literature is available to guide the development of multigrid solvers for nearly singular systems with the presence of a semi-definite term; see for instance (Lee et al., 2007; Schöberl, 1999a, 1999b). Particularly, Schöberl's seminal paper (Schöberl, 1999a) on the construction of parameter-robust multigrid schemes lists two requirements that must be satisfied for the top-left solve to be robust. The first requirement is a parameter-robust relaxation method; this is achieved by developing a space decomposition that stably captures the kernel of the semi-definite terms. The second requirement is a parameter-robust prolongation operator, i.e., one whose continuity constant is independent of the parameters. This is achieved by (approximately) mapping kernel functions on coarse grids to kernel functions on fine grids. We separately discuss both of these requirements below.

Table 3.1. The average number of FGMRES iterations per Newton iteration (total number of Newton iterations) for a nematic LC problem in a square slab. See the detailed problem description in Chapter 4.

					γ		
#refs	#dofs	10^1	10^2	10^3	10^4	10^5	10^6
1	5,340	33.75(4)	14.80(5)	6.20(5)	4.38(8)	7.18(11)	32.53(19)
2	21,080	75.00(5)	31.80(5)	11.60(5)	4.86(7)	5.83(12)	16.53(15)
3	83,760	>100	57.60(5)	24.60(5)	10.17(6)	46.75(8)	>100
4	333,920	>100	>100	90.80(5)	19.67(6)	>100	>100

In this chapter, we will construct a parameter-robust multigrid algorithm based on these works (Lee *et al.*, 2007; Schöberl, 1999a, 1999b). Some extensions of the analysis from Schöberl's work (Schöberl, 1999b) are given for the LC case. Then in order to verify the two aforementioned requirements of constructing the robust multigrid algorithm, a detailed example using the point-block Jacobi or star relaxation and the natural prolongation is illustrated for two-dimensional cholesteric problems.

3.1   Overview on Multigrid Methods

To solve minimising problems whose equilibrium equation turns out to be a large scale nonlinear system, *multigrid* (MG) methods have become the most useful and common choice for solving a large class of equations coming from various application, such as computational fluid dynamics, solid mechanics and molecular structures. Hence, we give an overview the construction of MG methods in this section and both linear and nonlinear systems are investigated to gain a better illustration. Readers may refer to Briggs *et al.* (2000) for a detailed introduction to MG and Henson (2003) a brief review of MG in solving particularly nonlinear equations.

3.1.1   *Solving linear equations*

To begin with, we consider a linear system with $\mathbf{u}, \mathbf{f} \in \mathbb{R}^n$ and $\mathbf{A} \in \mathbb{R}^{n \times n}$:

$$\mathbf{A}\mathbf{u} = \mathbf{f}, \tag{3.1}$$

where the solution $\mathbf{u}$ is sought. Theoretically, Gaussian elimination can be performed as a direct solver to seek the exact solution. However, this approach exhibits high computation costs and large memory storage for large scale problems. Therefore, it is more efficient to apply iterative methods to find an approximate solution $\mathbf{v}$ and the resulting error is simply given by

$$\mathbf{e} = \mathbf{u} - \mathbf{v}. \tag{3.2}$$

Remark 3.1. This error can be measured by the commonly used Euclidean norm or maximal norm, defined respectively by

$$\|\mathbf{e}\|_2 = \left\{ \sum_{j=1}^{n} e_j^2 \right\}^{1/2} \quad \text{and} \quad \|\mathbf{e}\|_\infty = \max_{1 \leq i \leq n} |e_j|.$$

Denote the residual $\mathbf{r}$ by

$$\mathbf{r} = \mathbf{f} - \mathbf{A}\mathbf{v}. \tag{3.3}$$

Since we are dealing with linear case, then combining these equations (3.1), (3.2), and (3.3), we deduce the following relation between error and residual:

$$\mathbf{A}\mathbf{e} = \mathbf{r}, \tag{3.4}$$

which is known as the *residual equation*.

There exists a wide variety of iterative methods and we start with the most classical ones, also called *smoothers*. They are the necessary building blocks for MG methods.

3.1.1.1 *Classical iterative methods*

The classical iterative methods start with a splitting procedure (Nocedal and Wright, 1999):

$$\mathbf{A} = \mathbf{M} - \mathbf{N}, \tag{3.5}$$

where it is supposed that $\mathbf{M}$ is nonsingular and its inverse is relatively easy to compute compared with $\mathbf{A}^{-1}$. The matrix $\mathbf{M}$ is called a *preconditioner* and should be a proper approximation of $\mathbf{A}$ in a sense that they are spectrally close. With this partition, we set up the iterative scheme as

$$\mathbf{u}_{k+1} = \mathbf{M}^{-1}\mathbf{N}\mathbf{u}_k + \mathbf{M}^{-1}\mathbf{f} \tag{3.6}$$

with a starting point $\mathbf{u}_0$ and it iterates until a satisfactory error tolerance is attained. By the classical convergence theory, the iteration (3.6) converges for any $\mathbf{f}$ and $\mathbf{u}_0$ if and only if the spectrum radius $\rho(\mathbf{M}^{-1}\mathbf{N}) < 1$.

Remark 3.2. The exact solution is actually the stationary point of the iteration (3.6) by letting $\mathbf{u}_{k+1} = \mathbf{u}_k = \mathbf{u}$. Then (3.6) becomes

$$\mathbf{Mu} = \mathbf{Nu} + \mathbf{f}$$
$$\Rightarrow \quad (\mathbf{M} - \mathbf{N})\mathbf{u} = \mathbf{f}$$
$$\Rightarrow \quad \mathbf{Au} = \mathbf{f}.$$

The most common partition used is

$$\mathbf{A} = \mathbf{D} - \mathbf{L} - \mathbf{U},$$

where $\mathbf{D}$ is the diagonal part and $\mathbf{L}, \mathbf{U}$ are the strictly lower and upper triangular parts, respectively. Choosing $\mathbf{M} = \mathbf{D}$, then $\mathbf{Au} = \mathbf{f}$ becomes

$$\mathbf{u} = \mathbf{D}^{-1}(\mathbf{L} + \mathbf{U})\mathbf{u} + \mathbf{D}^{-1}\mathbf{f},$$

and the scheme derived from this splitting is the so-called *Jacobi* method. We define its iteration matrix as

$$R_J = \mathbf{D}^{-1}(\mathbf{L} + \mathbf{U}) = \mathbf{I} - \mathbf{D}^{-1}\mathbf{A}.$$

If we choose $\mathbf{M} = \mathbf{D} - \mathbf{L}$, then we get the *Gauss–Seidel* method with its iteration matrix to be

$$R_{\mathrm{GS}} = (\mathbf{D} - \mathbf{L})^{-1}\mathbf{U}.$$

A modification of the splitting of $\mathbf{A}$ with a parameter ω gives

$$\omega\mathbf{A} = (\mathbf{D} - \omega\mathbf{L}) - (\omega\mathbf{U} + (1 - \omega)\mathbf{D}),$$

and we have the *Successive Over Relaxation* (SOR) scheme

$$\mathbf{u}^{n+1} = (\mathbf{D} - \omega\mathbf{L})^{-1}[\omega\mathbf{U} + (1 - \omega)\mathbf{D}]\mathbf{u}^n + (\mathbf{D} - \omega\mathbf{L})^{-1}\omega\mathbf{f},$$

with its iteration matrix

$$R_{\mathrm{SOR}} = \mathbf{I} - \left(\frac{\mathbf{D}}{\omega} - \mathbf{L}\right)^{-1}\mathbf{A}.$$

An important feature of these iterative methods is that they eliminate those high-frequency modes and leave behind the low-frequency modes, i.e., they smooth out the oscillatory parts (Briggs *et al.*, 2000). Due to this property, these schemes can be satisfyingly employed as the smoothing procedures in the later MG construction.

3.1.1.2 Two-grid schemes

So far, we have reviewed the basic iterative methods for solving linear systems. Now, we turn our concerns to the construction of MG schemes (see a schematic depiction of MG V-cycle in Falgout et $al.$ (2018) for instance), which is exactly the recursion of Two-$grid$ (TG) schemes. TG schemes start first with an iterative relaxation to smooth the error on the fine grid. Then a coarse-grid correction is applied by solving the residual equation on coarse grid.

Here, we use the notation of Ω^h, Ω^H for fine and coarse mesh, respectively. Take now step size $H = 2h$ for instance.

To solve the linear system (3.1) on the fine grid Ω^h, we perform a certain number of smoothing iterations and obtain an approximation solution $\mathbf{v}^h$ and its corresponding residual

$$\mathbf{r}^h = \mathbf{f}^h - \mathbf{A}^h \mathbf{v}^h,$$

and error

$$\mathbf{e}^h = \mathbf{u}^h - \mathbf{v}^h,$$

where $\mathbf{u}^h$ is the exact solution interpolated on fine grid Ω^h. But generally we do not have this exact solution in hand, thus the error is usually unknown. Instead, we get the residual equation

$$\mathbf{A}^h \mathbf{e}^h = \mathbf{r}^h. \tag{3.7}$$

Now we solve (3.7) on the coarse grid Ω^H to obtain the coarse-grid error $\mathbf{e}^H$, since it is much easier and quicker to solve problems on a coarse grid. This procedure is the so-called $coarse$-$grid$ $correction$ and it requires the $restriction$ operator I_h^H (from fine to coarse grid) and $prolongation$ operator I_H^h (from coarse to fine grid) between these two grids. The approximated error is then interpolated back into the fine grid to correct the fine-grid approximation solution:

$$\mathbf{v}^h \leftarrow \mathbf{v}^h + I_H^h \mathbf{e}^H.$$

The following procedure summarises the TG scheme:

- Pre-smooth ν_1 times on $\mathbf{A}^h \mathbf{u}^h = \mathbf{f}^h$ on fine grid Ω^h with an initial guess $\mathbf{v}^h$.
- Compute the residual $\mathbf{r}^h = \mathbf{f}^h - \mathbf{A}^h \mathbf{u}^h$ and then restrict it to the coarse grid by $\mathbf{r}^H = I_h^H \mathbf{r}^h$.

- Solve $\mathbf{A}^H \mathbf{e}^H = \mathbf{r}^H$ for the coarse-grid error $\mathbf{e}^H$ and interpolate it back to fine grid by $\mathbf{e}^h = I_H^h \mathbf{e}^H$.
- Correct the fine-grid approximation: $\mathbf{v}^h \leftarrow \mathbf{v}^h + \mathbf{e}^h$.
- Post-smooth ν_2 times on $\mathbf{A}^h \mathbf{u}^h = \mathbf{f}^h$ on fine grid Ω^h with initial guess $\mathbf{v}^h$ to obtain the final $\mathbf{v}^h$.

Combining the above procedures by matrix multiplication, we are able to derive the iteration matrix for TG scheme (see Lemma 1 in Borzí, 2002):

$$\mathbf{R}_{TG} = \mathbf{S}_h^{\nu_2}(\mathbf{I}^h - I_H^h(\mathbf{A}^H)^{-1} I_h^H \mathbf{A}^h)\mathbf{S}_h^{\nu_2},$$

where $\mathbf{S_h}$ denote the smoothing matrix and $\mathbf{I}^h$ the identity matrix on Ω^h.

3.1.1.3 *Multigrid schemes*

As TG schemes only tackle with the problem between two grid levels, it is natural to consider an adoption of multiple levels of grids, namely, $\{\Omega^{h_1}, \Omega^{h_2}, \Omega^{h_3}, \dots, \Omega^{h_k}\}$ with $h_1 > h_2 > \cdots > h_k$ and $h_{k-1} = 2h_k$. By recursively perform TG procedure, we derive the MG scheme. There are different ways of the recursion and thus lead to various type of MG schemes, but the essential ingredients of a MG scheme agree with that of a TG scheme, i.e., smoothing and coarse-grid correction.

Write I_k^{k-1} and I_{k-1}^k as the restriction and prolongation operator between grids Ω^{h_k} and $\Omega^{h_{k-1}}$, then the following concludes the MG scheme for $\mathbf{A}^k \mathbf{u}^k = \mathbf{f}^k$:

- If $k = 1$, then we can directly solve it by Gaussian elimination.
- Pre-smooth ν_1 times on $\mathbf{A}^k \mathbf{u}^k = \mathbf{f}^k$ on fine grid Ω^k.
- Compute the residual $\mathbf{r}^k = \mathbf{f}^k - \mathbf{A}^k \mathbf{u}^k$ and then restrict it to the coarse grid by $\mathbf{r}^{k-1} = I_k^{k-1} \mathbf{r}^k$.
- Set initial guess $\mathbf{e}^{k-1} = \mathbf{0}$.
- Use MG scheme γ times to solve $\mathbf{A}^{k-1} \mathbf{e}^{k-1} = \mathbf{r}^{k-1}$ for the coarse-grid error $\mathbf{e}^{k-1}$ and interpolate it back to fine grid by $\mathbf{e}^k = I_{k-1}^k \mathbf{e}^{k-1}$.
- Correct the fine-grid approximation: $\mathbf{v}^{h_k} \leftarrow \mathbf{v}^{h_k} + \mathbf{e}^{h_k}$.
- Post-smooth ν_2 times on $\mathbf{A}^{h_k} \mathbf{u}^{h_k} = \mathbf{f}^{h_k}$ on fine grid Ω^{h_k} to obtain the final $\mathbf{v}^{h_k}$.

The most commonly used and simplest version is the V-cycle, which is a direct extension of TG scheme by solving the coarse-grid residual equation using another TG process and proceeding to the next coarser grid until reaching at the coarsest grid. This is the case $\gamma = 1$ in the above procedures. See Figure 3.1 for a schematic description.

Another type of MG scheme is the W-cycle (i.e., $\gamma = 2$), where the step of coarse-grid correction is conducted twice at each level of grids. This is shown in Figure 3.2.

Additionally, if we construct the initial guess by interpolation from the solution of the coarsest grid equation, then *full MG* scheme is presented, which is denoted as F-cycle (see Figure 3.3).

The iteration matrices for V- and W-cycle can be found in Lemma 2 of Borzí (2002).

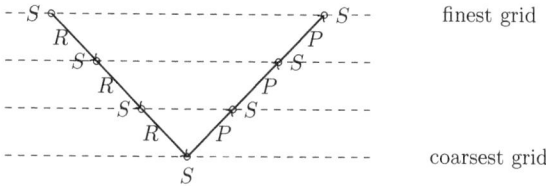

Fig. 3.1. V-cycle; S represents smoothing, R restriction of residual and P prolongation.

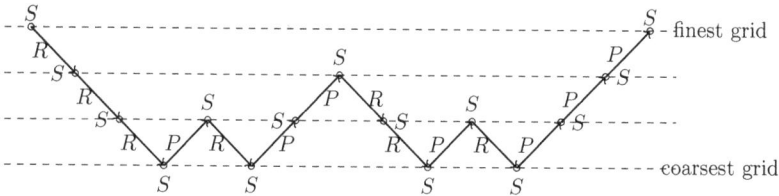

Fig. 3.2. W-cycle; S represents smoothing, R restriction of residual and P prolongation.

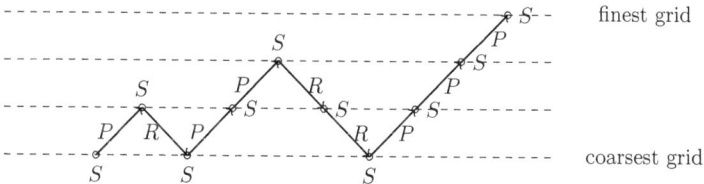

Fig. 3.3. F-cycle; S represents smoothing, R restriction of residual and P prolongation.

3.1.2   *Solving nonlinear equations*

Benefited from the linear case of MG, which provides an illustration on the underlying idea of MG schemes, we now consider the nonlinear case. Remembering that the key points of MG are smoothing and coarse-grid correction and it relies on solving the linear residual equation on the coarsest grid, we notice we need a different strategy for deriving the coarse-grid correction in the nonlinear case. There are two ways. The first one is using, for instance, Newton's method to linearise the original nonlinear problem, then apply the linear MG as derived in Section 3.1.1 for the Jacobian system in each Newton's iteration. We denote it as *Newton–MG*. The second approach is using the same idea of linear MG to deal with nonlinearity, where a full equation rather than the residual equation is solved on the coarse grid to correct the error, and this derives the so-called *Full Approximation Scheme* (FAS) but we will not discuss the details of FAS here.

Consider the nonlinear problem

$$A(\mathbf{u}) = \mathbf{f}, \tag{3.8}$$

where $\mathbf{u}, \mathbf{f} \in \mathbb{R}^n$ and $A(\cdot)$ is a nonlinear operator.

Similarly, we denote the error and residual, respectively, by

$$\mathbf{e} = \mathbf{u} - \mathbf{v} \quad \text{and} \quad \mathbf{r} = \mathbf{f} - A(\mathbf{v}),$$

with the approximate solution $\mathbf{v}$. Notice that since $A(\cdot)$ is nonlinear, we can not obtain the same residual equation (3.3) as in the linear case. Instead, we have

$$A(\mathbf{u}) - A(\mathbf{v}) = \mathbf{r}. \tag{3.9}$$

3.1.2.1   *Newton–MG scheme*

It is well-known that Newton's method is the most important method for solving nonlinear equations. In order to solve (3.8) by Newton's method, we first rewrite it as

$$F(\mathbf{u}) := A(\mathbf{u}) - \mathbf{f} = \mathbf{0},$$

and denote its Jacobian matrix by $J(\mathbf{u})$.

Since $\mathbf{u} = \mathbf{v} + \mathbf{e}$, we expand $F(\mathbf{u})$ at $\mathbf{v}$ to see

$$F(\mathbf{v} + \mathbf{e}) = F(\mathbf{v}) + J(\mathbf{v})\mathbf{e} + \text{higher order terms}$$

By neglecting the higher-order terms, we have a linearised version of the nonlinear problem:

$$J(\mathbf{v})\mathbf{e} = \mathbf{r},$$

where the MG scheme developed in Section 3.1.1 can be performed and we skip the details for brevity.

As we can see from the above discussion, Newton–MG is a combination of Newton's method as the outer iteration and MG as the inner iteration. It is noticeable that in some sense, Newton–MG does not employ the MG idea to the nonlinearity but applies Newton's method to treat the nonlinearity.

3.1.2.2 *Full Approximation Scheme*

Now we turn to dealing with nonlinearity directly under the framework of MG and this results in the so-called FAS. As described previously, MG scheme relies on two necessary steps, smoothing and coarse-grid correction. For the former step, the nonlinearity of the problem requires a nonlinear relaxation. In general, it is possible for us to find the corresponding nonlinear analogues of the classical iterative methods (introduced in Section 3.1.1.1) and thus deduce an approximation $\mathbf{v}^h$ to the fine-grid problem:

$$A^h(\mathbf{u}^h) = \mathbf{f}^h.$$

With the notation of error and residual, we presents the coarse-grid version of (3.9):

$$A^H(\mathbf{v}^H + \mathbf{e}^H) - A^H(\mathbf{v}^H) = \mathbf{r}^H, \tag{3.10}$$

where

$$\mathbf{v}^H = I_h^H \mathbf{v}^h \tag{3.11}$$

and the residual is

$$\mathbf{r}^H = I_h^H \mathbf{r}^h = I_h^H(\mathbf{f}^h - A^h(\mathbf{v}^h)). \tag{3.12}$$

Combining the above three equations yields

$$A^H(\mathbf{u}^H) := A^H(I_h^H \mathbf{v}^h + \mathbf{e}^H) = A^H(I_h^H \mathbf{v}^h) + I_h^H(\mathbf{f}^h - A^h(\mathbf{v}^h)) =: \mathbf{f}^H. \tag{3.13}$$

Remark 3.3. It is noticeable that the right-hand side of (3.13) is known. Thus, instead of having the residual equation (3.7) as in the linear case, we need to solve the full equation (3.13) on the coarse grid.

Suppose we have found the solution $\mathbf{u}^H$ of the coarse grid equation (3.13), then the coarse-grid error reduces to

$$\mathbf{e}^H = \mathbf{u}^H - I_h^H \mathbf{v}^h,$$

which will correct the fine grid approximation $\mathbf{v}^h$, i.e.,

$$\mathbf{v}^h \leftarrow \mathbf{v}^h + I_H^h \mathbf{e}^H.$$

Remark 3.4. In fact, one might think about directly using the interpolation of the coarse-grid solution $\mathbf{u}^H$ to fine grid as the correction. However, there is no information on the smoothness of the solution $\mathbf{u}^H$ and it generally contains oscillatory or high frequency components. Therefore, by interpolation from coarse grid to fine grid, we can get an inaccurate result. On the other hand, since the error usually contain low-frequency (smooth) data, it does not affect its accuracy by the introduction of interpolations. This explains why we employ the coarse-grid error to correct the fine-gird approximation, rather than a full approximation solution (one may refer to Chapter 3 of Briggs *et al.*, 2000).

The above TG version with the version of FAS for solving nonlinear equations is concluded as follows:

- Apply nonlinear pre-smoothing ν_1 times of $A^h(\mathbf{u}^h) = \mathbf{f}^h$ with initial guess $\mathbf{v}^h$ on fine grid Ω^h.
- Restrict the fine-grid approximation $\mathbf{v}^h$ and the residual by $\mathbf{v}^H = I_h^H \mathbf{v}^h$ and $\mathbf{r}^H = I_h^H(\mathbf{f}^h - A^h(\mathbf{v}^h))$, respectively.
- Seek the solution $\mathbf{u}^H$ of the coarse-grid full equation $A^H(\mathbf{u}^H) - A^H(\mathbf{v}^H) = \mathbf{r}^H$.
- Compute the coarse-grid error: $\mathbf{e}^H = \mathbf{u}^H - \mathbf{v}^H$.
- Interpolate the coarse-grid error back upon fine grid: $\mathbf{e}^h = I_H^h \mathbf{e}^H$.
- Correct the current fine-grid approximation by $\mathbf{v}^h \leftarrow \mathbf{v}^h + \mathbf{e}^h$.
- Apply nonlinear post-smoothing ν_2 times with initial guess $\mathbf{v}^h$ to get the final approximation $\mathbf{v}^h$.

By recursively performing the TG version of FAS discussed above, we derive the MG version of FAS for nonlinear equations. Similarly to the linear case, we also have the V-cycle, W-cycle and F-cycle as depicted in Figures 3.1–3.3.

Remark 3.5. If the Newton–MG is properly tuned with suitable inner iterations, then it can exhibit similar efficiency to that of FAS (Henson, 2003).

To illustrate the construction of a robust multigrid algorithm for LC problems using the idea from Schöberl (1999b), we discuss the combinations of the natural prolongation and two different types of relaxations for equal-constant nematic problems. Their satisfactions of the conditions (i.e., a kernel capturing relaxation and a robust prolongation) are checked. The numerical verifications are shown in Chapter 4.

For ease of notation, we consider the two-grid method applied to the equal-constant nematic case, and use subscripts h and H to distinguish fine and coarse mesh levels, respectively. That is to say, V_H represents the coarse-grid function space and we denote the associated operator $A_{H,\gamma}$: $V_H \to V_H^*$

$$(A_{H,\gamma}\mathbf{u}_H, \mathbf{v}_H)_0 := \mathfrak{a}^m(\mathbf{u}_H, \mathbf{v}_H)$$

with approximations $\mathbf{u}_H, \mathbf{v}_H$ on V_H. The analysis in this chapter can be extended to more complicated cases, e.g., with non-equal constants and more than two levels of grids.

For the domain Ω, we consider a non-overlapping triangulation $\mathcal{T}_H$, i.e.,

$$\cup_{T \in \mathcal{T}_H} T = \bar{\Omega} \text{ and } \mathrm{int}(T_i) \cap \mathrm{int}(T_j) = \emptyset \quad \forall T_i \neq T_j, \ T_i, T_j \in \mathcal{T}_H.$$

The fine grid $\mathcal{T}_h$ with $h = H/2$ is obtained by a regular refinement of the simplices in $\mathcal{T}_H$. In what follows we consider both the $[\mathbb{P}_1]^d$–$\mathbb{P}_1$ and $[\mathbb{P}_2]^d$–$\mathbb{P}_1$ discretisations.

3.2   Relaxation

After applying the AL method introduced in Section 2.3.1, the discrete linear variational form corresponding to the top-left block $\mathbf{A}_\gamma = \mathbf{A} + \gamma \mathbf{A}_*$ is given by

$$\mathfrak{a}^m(\mathbf{u}_h, \mathbf{v}_h) = K_c\, (\nabla \mathbf{u}_h, \nabla \mathbf{v}_h)_0 + 2\, (\lambda_j, \mathbf{u}_h \cdot \mathbf{v}_h)_0 + 4\gamma\, (\mathbf{n}_j \cdot \mathbf{u}_h, \mathbf{n}_j \cdot \mathbf{v}_h)_0 ,$$
(3.14)

with $\mathbf{u}_h \in V_h \subset \mathbf{H}_0^1(\Omega)$ being the trial function and $\mathbf{v}_h \in V_h$ the test function. Note that $\mathbf{n}_j$ and λ_j are the current approximations to the director $\mathbf{n}$ and the Lagrange multiplier λ, respectively, in the Newton iteration. The first two terms of $\mathfrak{a}^m$ are symmetric and coercive because of the running assumption of uniform non-negativity of λ_j. The kernel of the semi-definite term involving γ is

$$\mathcal{N}_h = \{\mathbf{u}_h \in V_h : \mathbf{n}_j \cdot \mathbf{u}_h = 0 \text{ a.e.}\}.$$
(3.15)

In the case of γ being very large, the variational problem involving (3.14) is nearly singular and common relaxation methods like Jacobi and Gauss–Seidel will not yield effective multigrid cycles, as we explain below.

Relaxation schemes can be devised in a generic way by considering *space decompositions*

$$V_h = \sum_{i=1}^{M} V_i, \tag{3.16}$$

where the sum of vector spaces on the right is not necessarily a direct sum [Xu (1992)]. For example, if $V_h = \operatorname{span}(\varphi_1, \ldots, \varphi_M)$, Jacobi and Gauss–Seidel iterations are induced by the space decomposition

$$V_i = \operatorname{span}(\varphi_i), \tag{3.17}$$

where the updates are performed additively for Jacobi and multiplicatively for Gauss–Seidel. One of the key insights of Lee *et al.* (2007) and Schöberl (1999a) was that the key requirement for parameter-robustness when applied to nearly singular problems is that the space decomposition must satisfy the *kernel-capturing property*

$$\mathcal{N}_h = \sum_{i=1}^{M} (V_i \cap \mathcal{N}_h), \tag{3.18}$$

that is, any kernel function can be written as a sum of kernel functions drawn from the subspaces. In particular, each subspace V_i must be rich enough to support kernel functions; in our context, this is not satisfied by the choice (3.17), accounting for its poor behaviour shown in Table 3.1 as $\gamma \to \infty$.

In the mesh triangulation $\mathcal{T}_h$, we denote the *star* of a vertex v_i as the patch of elements sharing v_i, i.e.,

$$\operatorname{star}(v_i) := \bigcup_{T \in \mathcal{T}_h : v_i \in T} T.$$

This induces an associated space decomposition, called the *star patch*, by

$$V_i := \{ \mathbf{u}_h \in V_h : \operatorname{supp}(\mathbf{u}_h) \subset \operatorname{star}(v_i) \}.$$

This is illustrated in Figure 3.4(a). We call the induced relaxation method a *star iteration*. In effect, each subspace solves for the degrees of freedom in the interior of the patch of cells, with homogeneous Dirichlet conditions on the boundary of the patch. Given a vertex or edge midpoint v_i,

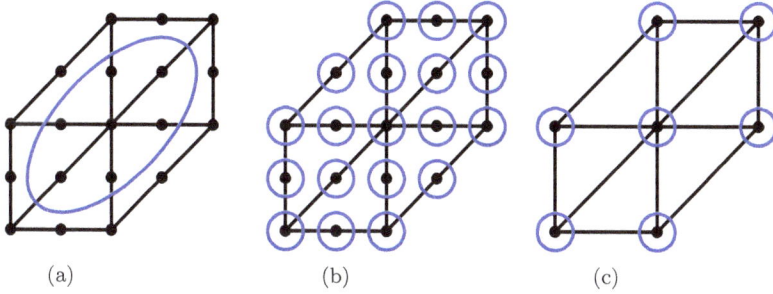

Fig. 3.4. Illustrations of the star patch of the center vertex (a) and the point-block patch (b) for the finite element pair $[\mathbb{P}_2]^2$-$\mathbb{P}_1$. Note that these two patches are the same for $[\mathbb{P}_1]^2$-$\mathbb{P}_1$ discretisation (c). Here, black dots represent the degrees of freedom, and the blue lines gather degrees of freedom solved for simultaneously in the relaxation.

we denote the *point-block* patch V_i as the span of the basis functions associated with degrees of freedom that evaluate a function at v_i (see Figure 3.4(b)). The induced relaxation method solves for all colocated degrees of freedom simultaneously. These two space decompositions coincide for the $[\mathbb{P}_1]^d$-$\mathbb{P}_1$ discretisation (see Figure 3.4(c)).

We now briefly explain why these two decompositions approximately satisfy the kernel-capturing condition (3.18) for the finite element pair $[\mathbb{P}_1]^d$-$\mathbb{P}_1$. First, we define an approximate kernel

$$\tilde{\mathcal{N}}_h = \{\mathbf{u}_h \in V_h : \mathbf{n}_j \cdot \mathbf{u}_h = 0 \text{ on each vertex}\}. \tag{3.19}$$

Since $\mathbf{n}_j$ is the current approximation to the director $\mathbf{n}$, we have $\mathbf{n}_j \in V_h = \sum_i V_i$. We are therefore able to express $\mathbf{n}_j$ as $\mathbf{n}_j = \sum_i \mathbf{n}_j^i$, where $\mathbf{n}_j^i \in V_i$ describes the function at the vertex v_i. Similarly, we split $\mathbf{u}_h$ into $\mathbf{u}_h = \sum_i \mathbf{u}_h^i$ with $\mathbf{u}_h^i \in V_i$. For each vertex v_i, the requirement $\mathbf{u}_h \in \tilde{\mathcal{N}}_h$ yields

$$\mathbf{n}_j^i \cdot \mathbf{u}_h^i = 0 \quad \forall i. \tag{3.20}$$

The definition of V_i ensures that $\mathbf{u}_h^i$ and $\mathbf{n}_j^i$ are only supported on the interior of the star of v_i. We deduce that on each vertex

$$\mathbf{n}_j^k \cdot \mathbf{u}_h^i = 0 \quad \forall i \neq k,$$

which yields $\sum_k \mathbf{n}_j^k \cdot \mathbf{u}_h^i = \mathbf{n}_j \cdot \mathbf{u}_h^i = 0$. Hence, $\mathbf{u}_h^i \in \tilde{\mathcal{N}}_h \forall i$ and we obtain the kernel-capturing condition (3.18) for the approximate kernel $\tilde{\mathcal{N}}_h$.

For the $[\mathbb{P}_2]^d$-$\mathbb{P}_1$ finite element pair, the satisfaction of the kernel-capturing property for the approximate kernel follows along similar lines.

For the point-block patch, (3.20) still holds. The star patch uses larger subspaces, each one including multiple point-block patches, but it can be easily verified that (3.20) is still fulfilled.

3.2.1 *Robustness analysis of the approximate kernel*

While we are not able to prove the kernel capturing property for the exact kernel (3.15), we can still obtain the spectral inequalities

$$c_1 D_{h,\gamma} \le A_{h,\gamma} \le c_2 D_{h,\gamma}, \tag{3.21}$$

when using the approximate kernel (3.19). Here, $D_{h,\gamma}$ is the preconditioner to be specified later for the operator $A_{h,\gamma}$ and $C \le D$ represents $\|\mathbf{u}\|_C \le \|\mathbf{u}\|_D$ for all $\mathbf{u}$. We prove that c_1 depends on γ, but the dependence can be well controlled so that the preconditioner is not badly affected by varying γ, while c_2 is always independent of γ. For simplicity, we prove the case for the equal-constant nematic case with the $[\mathbb{P}_1]^d$-$\mathbb{P}_1$ discretisation; extensions to the nonequal-constant cholesteric case and to the $[\mathbb{P}_2]^d$-$\mathbb{P}_1$ discretisation are possible.

We define the operator associated to $\mathfrak{a}^m$, $A_{h,\gamma} : V_h \to V_h^*$, by

$$(A_{h,\gamma}\mathbf{u}_h, \mathbf{v}_h)_0 := \mathfrak{a}^m(\mathbf{u}_h, \mathbf{v}_h).$$

For the space decomposition $V_h = \sum_i V_i$, we denote the lifting operator (the natural inclusion) by $I_i : V_i \to V_h$ and choose the Galerkin subspace operator $A_i : V_i \to V_i$ to satisfy

$$(A_i\mathbf{u}_i, \mathbf{v}_i)_0 := (A_{h,\gamma}I_i\mathbf{u}_i, I_i\mathbf{v}_i)_0 \quad \forall \mathbf{u}_i, \mathbf{v}_i \in V_i.$$

This implies that $A_i = I_i^* A_{h,\gamma} I_i$.

The additive Schwarz preconditioner $D_{h,\gamma}$ for a problem $A_{h,\gamma}w_h = d_h$ associated with the space decomposition (3.16) is defined by the action of its inverse (Xu, 1992):

$$w_h = D_{h,\gamma}^{-1} d_h$$

given by

$$w_h = \sum_{i=1}^{M} I_i w_i,$$

with $w_i \in V_i$ being the unique solution of

$$(A_i w_i, v_i)_0 = (d_h, I_i v_i)_0 \quad \forall v_i \in V_i.$$

Hence, we can rewrite the preconditioning operator $D_{h,\gamma}^{-1}$ in operator form as

$$D_{h,\gamma}^{-1} = \sum_{i=1}^{M} I_i A_i^{-1} I_i^*.$$

We now state for completeness a classical result in the analysis of additive Schwarz preconditioners, see, e.g., Schöberl (1999b, Theorem 3.1) and the references therein.

Theorem 3.1. *Define the splitting norm for* $\mathbf{u}_h \in V_h$ *as*

$$|||\mathbf{u}_h|||^2 := \inf_{\substack{\mathbf{u}_h = \sum_i I_i \mathbf{u}_i \\ \mathbf{u}_i \in V_i}} \sum_{i=1}^{M} \|\mathbf{u}_i\|_{A_i}^2.$$

This splitting norm is equal to the norm $\|\mathbf{u}_h\|_{D_{h,\gamma}} := (D_{h,\gamma}\mathbf{u}_h, \mathbf{u}_h)_0^{1/2}$ *generated by the additive Schwarz preconditioner, i.e., it holds that*

$$|||\mathbf{u}_h|||^2 = \|\mathbf{u}_h\|_{D_{h,\gamma}}^2 \quad \forall \mathbf{u}_h \in V_h.$$

To build intuition, let us examine why Jacobi relaxation defined by the space decomposition (3.17) is not robust as $\gamma \to \infty$. With (3.17), the decomposition $\mathbf{u}_h = \sum_i^M \mathbf{u}_i, \mathbf{u}_i \in V_i$ is unique. It yields that

$$\|\mathbf{u}_h\|_{D_{h,\gamma}}^2 = |||\mathbf{u}_h|||^2 = \sum_i (A_i \mathbf{u}_i, \mathbf{u}_i)_0 = \sum_i (A_{h,\gamma}\mathbf{u}_i, \mathbf{u}_i)_0$$

$$\lesssim (1+\gamma) \sum_i \|\mathbf{u}_i\|_1^2 \lesssim \frac{1+\gamma}{h^2} \sum_i \|\mathbf{u}_i\|_0^2 \lesssim \frac{1+\gamma}{h^2} \|\mathbf{u}_h\|_0^2$$

$$\lesssim \frac{1+\gamma}{h^2} \|\mathbf{u}_h\|_{A_{h,\gamma}}^2. \tag{3.22}$$

Note that the bound in (3.22) is parameter-dependent and deteriorates as $\gamma \to \infty$ or $h \to 0$.

In order to deduce the robustness result for our approximate kernel (3.19), we first derive the following lemma.

Lemma 3.2. *Let* $\mathbf{u}_0 = \sum_i^M \mathbf{u}_0^i \in \tilde{\mathcal{N}}_h$ *and assume* $\mathbf{n}_j \in [\mathbb{P}_1]^d$. *Then it holds that*

$$\sum_i^M \|\mathbf{u}_0^i \cdot \mathbf{n}_j\|_{L^2(\Omega)}^2 \lesssim h^2 \|\mathcal{D}\mathbf{n}_j\|_{L^\infty(\Omega)}^2 \|\mathbf{u}_0\|_{L^2(\Omega)}^2,$$

where $\mathcal{D}\mathbf{n}_j$ *denotes the Jacobian matrix of* $\mathbf{n}_j$.

Proof. Consider the vertex v_i on the boundary of an element T. As $\mathbf{n}_j \in [\mathbb{P}_1]^d$, we have

$$(\mathbf{u}_0^i \cdot \mathbf{n}_j)(\mathbf{x}) = \mathbf{u}_0^i(\mathbf{x}) \cdot \mathbf{n}_j(v_i) + \mathbf{u}_0^i(\mathbf{x}) \cdot [\mathcal{D}\mathbf{n}_j(v_i)(\mathbf{x} - v_i)] \quad \forall \mathbf{x} \in T.$$

Note that $\mathbf{u}_0^i \cdot \mathbf{n}_j$ vanishes at the vertex v_i as $\mathbf{u}_0 \in \tilde{\mathcal{N}}_h$. Moreover, we know $\mathbf{u}_0^i(\mathbf{x})/\|\mathbf{u}_0^i(\mathbf{x})\|$ is constant on the interior of the patch around v_i, and $\mathbf{u}_0^i(\mathbf{x})$ is zero on the boundary of the patch, since we can write $\mathbf{u}_0^i(\mathbf{x}) = \mathbf{u}_0(v_i)\psi_i(\mathbf{x})$ with ψ_i denoting the scalar piecewise linear basis function (vanishing outside the patch) associated with v_i. Therefore, we can deduce $\mathbf{u}_0^i(\mathbf{x}) \cdot \mathbf{n}_j(v_i) = 0$ on T. In addition, we have $\|\mathbf{x} - v_i\| \lesssim h$ on the element T. We thus conclude that

$$\|\mathbf{u}_0^i \cdot \mathbf{n}_j\|_{L^2(T)} \lesssim h\|\mathcal{D}\mathbf{n}_j\|_{L^\infty(T)}\|\mathbf{u}_0^i\|_{L^2(T)}.$$

From this we are able to show that for both the star and point-block patches around v_i,

$$\sum_i \|\mathbf{u}_0^i \cdot \mathbf{n}_j\|_{L^2(\mathrm{patch}(v_i))}^2 \lesssim \sum_i h^2\|\mathcal{D}\mathbf{n}_j\|_{L^\infty(\mathrm{patch}(v_i))}^2\|\mathbf{u}_0^i\|_{L^2(\mathrm{patch}(v_i))}^2$$

$$\lesssim h^2\|\mathcal{D}\mathbf{n}_j\|_{L^\infty(\Omega)}^2 \sum_i \|\mathbf{u}_0^i\|_{L^2(\Omega)}^2$$

$$\lesssim h^2\|\mathcal{D}\mathbf{n}_j\|_{L^\infty(\Omega)}^2\|\mathbf{u}_0\|_{L^2(\Omega)}^2.$$

Therefore, with the local support of $\mathbf{u}_0^i$ we have

$$\sum_i \|\mathbf{u}_0^i \cdot \mathbf{n}_j\|_{L^2(\Omega)}^2 = \sum_i \|\mathbf{u}_0^i \cdot \mathbf{n}_j\|_{L^2(\mathrm{patch}(v_i))}^2 \lesssim h^2\|\mathcal{D}\mathbf{n}_j\|_{L^\infty(\Omega)}^2\|\mathbf{u}_0\|_{L^2(\Omega)}^2. \qquad \square$$

We now derive the general form of the spectral bounds in (3.21). This follows a similar approach to Schöberl (1999b, Theorem 4.1), but with a different assumption on the splitting approximation, to allow for a dependence on γ. Given a space decomposition $V_h = \sum_i^M V_i$, we define its *overlap* N_O as

$$N_O := \max_{1 \le i \le M} \sum_{j=1}^M g_{ij},$$

where

$$g_{ij} = \begin{cases} 1 & \text{if } \exists \mathbf{v}_i \in V_i, \mathbf{v}_j \in V_j : |\mathrm{supp}(\mathbf{v}_i) \cap \mathrm{supp}(\mathbf{v}_j)| > 0, \\ 0 & \text{otherwise} \end{cases}$$

measures the interaction between each subspace.

Theorem 3.3. *Let $\{V_i\}$ be a subspace decomposition of V_h with overlap N_O. Assume that the finite element pair V_h-Q_h for (u, λ) is inf-sup stable for the mixed problem*

$$\mathcal{B}((\mathbf{u}, \lambda); (\mathbf{v}, \mu)) := K_c (\nabla \mathbf{u}, \nabla \mathbf{v})_0 + 2 (\lambda, \mathbf{n}_j \cdot \mathbf{v})_0 + 2 (\mu, \mathbf{n}_j \cdot \mathbf{u})_0$$
$$= \mathcal{F}(\mathbf{v}, \mu) \quad \forall (\mathbf{v}, \mu) \in V_h \times Q_h,$$

where $\mathcal{F}$ is a known functional. Furthermore, assume that the function $\mathbf{u}_h \in V_h$ and the kernel function $\mathbf{u}_0 \in \mathcal{N}_h$ can be split locally with estimates depending on the mesh size h and possibly on γ if the kernel-capturing property is not satisfied:

$$\inf_{\substack{\mathbf{u}_h = \sum_i \mathbf{u}_h^i \\ \mathbf{u}_h^i \in V_i}} \sum_i \|\mathbf{u}_h^i\|_1^2 \leq c_1(h) \|\mathbf{u}_h\|_0^2,$$

$$\inf_{\substack{\mathbf{u}_0 = \sum_i \mathbf{u}_0^i \\ \mathbf{u}_0^i \in V_i}} \sum_i \|\mathbf{u}_0^i\|_{A_{h,\gamma}}^2 \leq (c_2(h) + c_3(h, \gamma)) \|\mathbf{u}_0\|_0^2.$$

Then the additive Schwarz preconditioner $D_{h,\gamma}$ built on the decomposition $\{V_i\}$ satisfies

$$(c_1(h) + c_2(h) + c_3(h, \gamma))^{-1} D_{h,\gamma} \leq A_{h,\gamma} \leq N_O D_{h,\gamma}, \tag{3.23}$$

with constants c_1 and c_2 independent of γ.

Proof. The upper bound can be directly given by Schöberl (1999b, Lemma 3.2) independent of the form of partial differential equations.

For the lower bound, choose $\mathbf{u}_h \in V_h$ and split it into $\mathbf{u}_h = \mathbf{u}_0 + \mathbf{u}_1$, by solving

$$\mathcal{B}((\mathbf{u}_1, \lambda_1), (\mathbf{v}_h, \mu_h)) = 2 (\mu_h, \mathbf{n}_j \cdot \mathbf{u}_h)_0 \quad \forall (\mathbf{v}_h, \mu_h) \in V_h \times Q_h. \tag{3.24}$$

Testing with $\mathbf{v}_h = 0$ in (3.24), we obtain that

$$(\mu_h, \mathbf{n}_j \cdot \mathbf{u}_1)_0 = (\mu_h, \mathbf{n}_j \cdot \mathbf{u}_h)_0 \quad \forall \mu_h \in Q_h.$$

Furthermore, since the current approximation $\mathbf{n}_j$ is well-controlled as from Assumption 2.3, $\mathbf{n}_j \cdot \mathbf{u}$ belongs to $L^2(\Omega)$. Hence, $\mathbf{n}_j \cdot \mathbf{u}_0 = 0$ a.e., that is to say $\mathbf{u}_0 \in \mathcal{N}_h$.

By stability of the finite element pair V_h-Q_h, we have

$$\|\mathbf{u}_1\|_1 \lesssim \sup_{\substack{\mathbf{v}_h \in V_h \\ \mu_h \in Q_h}} \frac{\mathcal{B}((\mathbf{u}_1, \lambda_1), (\mathbf{v}_h, \mu_h))}{\|(\mathbf{v}_h, \mu_h)\|}$$

$$\lesssim \sup_{\substack{\mathbf{v}_h \in V_h \\ \mu_h \in Q_h}} \frac{\|\mathbf{n}_j \cdot \mathbf{u}_h\|_0 \|\mu_h\|_0}{\|(\mathbf{v}_h, \mu_h)\|}$$

$$\leq \|\mathbf{n}_j \cdot \mathbf{u}_h\|_0.$$

It follows that

$$\|\mathbf{u}_1\|_1 \lesssim \|\mathbf{u}_h\|_0$$

by the boundedness of $\mathbf{n}_j$ and

$$\|\mathbf{u}_1\|_1 \lesssim \gamma^{-1/2} \|\mathbf{u}_h\|_{A_{h,\gamma}}$$

by the form of the operator $A_{h,\gamma}$, respectively. Using $\mathbf{u}_0 = \mathbf{u}_h - \mathbf{u}_1$, we have in addition that

$$\|\mathbf{u}_0\|_1 \lesssim \|\mathbf{u}_h\|_1.$$

We now calculate

$$\|\mathbf{u}_h\|_{D_{h,\gamma}}^2 = \|\|\mathbf{u}_h\|\|^2$$

$$\leq \inf_{\substack{\mathbf{u}_1 = \sum_i \mathbf{u}_1^i \\ \mathbf{u}_1^i \in V_i}} \sum_i \|\mathbf{u}_1^i\|_{A_{h,\gamma}}^2 + \inf_{\substack{\mathbf{u}_0 = \sum_i \mathbf{u}_0^i \\ \mathbf{u}_0^i \in V_i}} \sum_i \|\mathbf{u}_0^i\|_{A_{h,\gamma}}^2$$

$$\lesssim (1 + \gamma) \inf_{\substack{\mathbf{u}_1 = \sum_i \mathbf{u}_1^i \\ \mathbf{u}_1^i \in V_i}} \sum_i \|\mathbf{u}_1^i\|_1^2 + (c_2(h) + c_3(h, \gamma)) \|\mathbf{u}_0\|_0^2$$

$$\lesssim (1 + \gamma) c_1(h) \|\mathbf{u}_1\|_0^2 + (c_2(h) + c_3(h, \gamma)) \|\mathbf{u}_0\|_1^2$$

$$\lesssim (1 + \gamma) c_1(h) \|\mathbf{u}_1\|_1^2 + (c_2(h) + c_3(h, \gamma)) \|\mathbf{u}_h\|_1^2$$

$$\lesssim (c_1(h) + c_2(h) + c_3(h, \gamma)) \|\mathbf{u}_h\|_{A_{h,\gamma}}^2, \tag{3.25}$$

completing the proof of the spectral estimates (3.23).   □

Remark 3.6. Note that in Theorem 3.3, if the kernel-capturing property (3.18) is satisfied, then c_3 will be zero. Hence, we will instead get a parameter-independent result.

Corollary 3.4. *In Theorem 3.3, if we take V_h-Q_h to be constructed by the $[\mathbb{P}_1]^d$-$\mathbb{P}_1$ element, it holds that*

$$\left(c_1(h) + c_2(h) + \gamma h^2 \|\mathcal{D}\mathbf{n}_j\|_\infty^2\right)^{-1} D_{h,\gamma} \leq A_{h,\gamma} \leq N_O D_{h,\gamma},$$

with constants $c_1(h)$, $c_2(h) \sim \mathcal{O}(h^{-2})$.

Proof. We follow the main argument of Theorem 3.3. We have only proven the kernel-capturing property for the approximate kernel (3.19) rather than (3.15), and need to account for this in the estimates. From Lemma 3.2 and the definition of $A_{h,\gamma}$ we have that

$$c_3(h,\gamma) = \gamma h^2 \|\mathcal{D}\mathbf{n}_j\|_\infty^2.$$

With the choice of $V_h = [\mathbb{P}_1]^d$, we will use the so-called *inverse inequality* (its proof can be found in any finite element book, e.g., Ciarlet, 1978) which states that

$$\|\mathbf{v}_h\|_1 \lesssim h^{-1} \|\mathbf{v}_h\|_0 \quad \forall \mathbf{v}_h \in V_h.$$

Therefore, it is straightforward to obtain that c_1 and c_2 are actually $\mathcal{O}(h^{-2})$. Notice here we have also used the form of $\|\cdot\|_{A_{h,\gamma}}$ in estimating $c_2(h)$.

Finally, substituting the form of c_3 in (3.25), we derive

$$\|\mathbf{u}_h\|_{D_{h,\gamma}}^2 \lesssim \left(c_1(h) + c_2(h) + \gamma h^2 \|\mathcal{D}\mathbf{n}_j\|_\infty^2\right) \|\mathbf{u}_h\|_{A_{h,\gamma}}^2,$$

with constants $c_1(h)$, $c_2(h) \sim \mathcal{O}(h^{-2})$. $\qquad\qquad\square$

Corollary 3.4 implies that we cannot entirely get rid of parameter γ in the spectral estimates if the kernel-capturing property for the kernel (3.15) is not satisfied and instead we get an additional factor of $\gamma h^2 \|\mathcal{D}\mathbf{n}_j\|_\infty^2$. However, this γ-dependence can be well controlled and does not impinge on the effectiveness of our smoother; the dependence improves as the mesh becomes finer or as $\mathbf{n}_j$ becomes smoother.

3.3 Prolongation

To construct a parameter-robust multigrid method, the prolongation operator is also required to be continuous (in the energy norm associated with the PDE) with the continuity constant independent of the penalty parameter γ (Schöberl, 1999b, Theorem 4.2). In the context of the Oseen, Navier–Stokes, and linear elasticity equations, the prolongation operator was modified in order to guarantee that the continuity constant is γ-independent

(Benzi and Olshanskii, 2006; Farrell *et al.*, 2019; Schöberl, 1999b). However, in our experiments with the Oseen–Frank system, we observe robust convergence with respect to γ, even when using the (cheaper) standard prolongation. This can be seen in Tables 4.7 and 4.8 of Chapter 4, for example. Hence, we will use the standard prolongation with no modification in this part of work.

Remark 3.7. Since both discretisations $[\mathbb{P}_1]^d$-$\mathbb{P}_1$ and $[\mathbb{P}_2]^d$-$\mathbb{P}_1$ are nested, i.e., $V_H \subset V_h$, the standard prolongation is actually a continuous (in the H^1-norm) natural inclusion.

3.4 Summary

In this chapter, we discussed constructing a robust multigrid algorithm for solving the augmented top-left block in the derived saddle point system (2.19) of LC problems. Two essential ingredients for the guarantee of robustness were examined: a relaxation that captures the kernel of the augmentation term and a prolongation operator that possesses a parameter-independent continuity constant. We will present some numerical results to verify the effectiveness of our constructed AL preconditioner in the next chapter.

Chapter 4

Numerical Experiments for Nematics and Cholesterics

4.1 Algorithm Details

In the following numerical experiments, we use the $[\mathbb{P}_2]^3$–$\mathbb{P}_1$ element pair and use flexible GMRES (Saad, 1993) as the outermost linear solver, since GMRES (Saad and Schultz, 1986) is applied in the multigrid relaxation. An absolute tolerance of 10^{-8} was used for the nonlinear solver, except for the convergence rate tests in Figure 4.4, which used 10^{-10}. A relative tolerance of 10^{-4} was used for the inner linear solver. We use the full block factorisation preconditioner

$$
\mathcal{Q}^{-1} = \begin{bmatrix} \mathbf{I} & -\tilde{\mathbf{A}}_\gamma^{-1}\mathbf{B}^\top \\ \mathbf{0} & \mathbf{I} \end{bmatrix} \begin{bmatrix} \tilde{\mathbf{A}}_\gamma^{-1} & \mathbf{0} \\ \mathbf{0} & \tilde{\mathbf{S}}_\gamma^{-1} \end{bmatrix} \begin{bmatrix} \mathbf{I} & \mathbf{0} \\ -\mathbf{B}\tilde{\mathbf{A}}_\gamma^{-1} & \mathbf{I} \end{bmatrix},
$$

where $\mathbf{I}$ is the identity matrix and $\tilde{\mathbf{A}}_\gamma^{-1}$ represents solving the top-left block $\mathbf{A}_\gamma$ inexactly by our specialised multigrid algorithm described in the previous chapter and the Schur complement approximation $\tilde{\mathbf{S}}_\gamma^{-1}$ is given by (2.27). The multiplier mass matrix inverse $\mathbf{M}_\lambda^{-1}$ is solved using Cholesky factorisation.

For $\tilde{\mathbf{A}}_\gamma^{-1}$, we perform a multigrid V-cycle, where the problem on the coarsest grid is solved exactly by Cholesky decomposition. On each finer level, as relaxation we perform three GMRES iterations preconditioned by the additive star (denoted as ALMG-STAR) iteration or additive point-block Jacobi (denoted as ALMG-PBJ) iteration. In order to achieve convergence results independent of the number of cores used in parallel, we only report iteration counts using additive relaxation, although multiplicative ones generally give better convergence. The star and Vanka relaxation methods are implemented using the PCPATCH preconditioner recently included in PETSc (Farrell *et al.*, 2021).

Code availability: For reproducibility, both the solver code (Xia, 2020) and the exact version of Firedrake (Firedrake-Zenodo, 2020) used to produce the numerical results of this chapter have been archived on Zenodo. An installation of Firedrake with components matching those used in this chapter can be obtained by following the instructions at https://www.fire drakeproject.org/download.html with

```
python3 firedrake-install --doi 10.5281/zenodo.4249051.
```

4.1.1   *Newton–Kantorovich method*

It is well known that Newton's method is the most useful and widely used numerical approach when encountering nonlinearity in finding a root of a problem. There is a rich history of its development and readers should refer to Birkisson (2013) and Deuflhard (2011). We should note that the original Newton's method (i.e., Newton–Raphson) is generalised to function spaces by (Kantorovich, 1948) and this version is called Newton–Kantorovich method.

For a nonlinear problem

$$F(x) = 0,$$

with $F : X \subset Y \to Z$ where X is an open subset of Y and both Y and Z are Banach spaces, the Newton iteration is written as

$$x_{k+1} = x_k - F'(x_k)^{-1} F(x_k),$$

where $F'(x_k)$ is the Frechét derivative of the operator F at x_k with an initial guess x_0. In practice, the solution of

$$F'(x_k)\delta_k = -F(x_k) \tag{4.1}$$

is sought first and then we update the numerical solution by $x_k \leftarrow x_k + \delta_k$. Hence, the underlying idea of Newton's method is actually sequentially solving the linearised version of the nonlinear problem.

We refer to Section 2.3 in Birkisson (2013) for the classical convergence theory of Newton's method in function spaces. In our implementation, we employ this method to solve nonlinear system while utilising a multigrid method to precondition the linearised problem (4.1).

4.2   Numerical Results

We denote #refs and #dofs as the number of mesh refinements and degrees of freedom, respectively, in the following experiments. The test problems in this section assume that the domain represents a uniform slab in the xy-plane, i.e., $\mathbf{n}$ may have a nonzero z-component, but $\frac{\partial \mathbf{n}}{\partial z} = \mathbf{0}$. Hence, though the domain is in two dimensions, we use the Cartesian representation of the director $\mathbf{n} = (n_x, n_y, n_z)$ throughout this chapter.

4.2.1   *Periodic boundary condition in a square slab*

Following the nematic benchmarks in (Adler *et al.*, 2016, Section 5.1), we consider a generalised twist equilibrium configuration in a square $\Omega = [0,1] \times [0,1]$, which has an analytical solution (Stewart, 2004). We will investigate the robustness of the solver when applied to unequal Frank constants and nonzero cholesteric pitch.

We impose periodic boundary conditions in the x-direction and Dirichlet boundary conditions in the y-direction, with values

$$\mathbf{n} = [\cos \vartheta_0, 0, -\sin \vartheta_0]^\top \quad \text{on} \quad y = 0,$$
$$\mathbf{n} = [\cos \vartheta_0, 0, \sin \vartheta_0]^\top \quad \text{on} \quad y = 1,$$

where $\vartheta_0 = \pi/8$.

We first consider parameter values $K_1 = 1.0$, $K_2 = 1.2$, $K_3 = 1.0$, $q_0 = 0$ when solving the minimisation problem (2.1). The exact solution is given by

$$\mathbf{n} = [\cos(\vartheta_0(2y-1)), 0, \sin(\vartheta_0(2y-1))]^\top,$$

with true free energy $2K_2\vartheta_0^2 \approx 0.37011$. An example of the pure twist configuration is illustrated in Figure 4.1.

We use an initial guess of $\mathbf{n}_0 = [1, 0, 0]^\top$ in the Newton iteration and a 10×10 mesh of triangles of negative slope as the coarse grid.

We first compare in Table 4.1 the nonlinear convergence of the Newton linearisation (2.16) against that of the Picard iteration (2.17) we propose. For these experiments, we use the augmented Lagrangian preconditioner with ideal inner solvers (denoted as ALLU), i.e., where the top-left block is solved exactly by LU factorisation. The Picard iteration requires substantially fewer nonlinear iterations for large γ. We expect that this relates to the degradation of the coercivity estimate given in Lemma 2.5. Similar

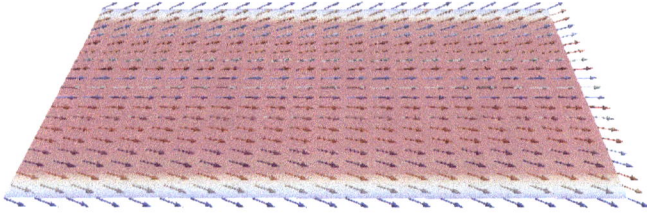

Fig. 4.1. A sample solution of the twist configuration. Colours represent the magnitude of directors.

Table 4.1. A comparison of the nonlinear convergence of the Newton linearisation (2.16) and the Picard iteration (2.17) using ideal inner solvers for a nematic LC problem in a square slab. The table shows the average number of outer FGMRES iterations per nonlinear iteration and the total nonlinear iterations in brackets.

	#refs	#dofs	10^3	10^4	10^5	10^6
					γ	
Newton	1	5,340	2.20 (5)	1.14 (7)	1.00 (10)	1.00 (19)
	2	21,080	3.20 (5)	1.14 (7)	1.00 (12)	1.00 (15)
	3	83,760	3.83 (6)	1.57 (7)	1.11 (9)	1.00 (14)
	4	333,920	4.67 (6)	2.14 (7)	1.00 (7)	1.00 (11)
	5	1,333,440	5.17 (6)	2.43 (7)	1.57 (7)	1.00 (10)
Picard	1	5,340	2.00 (5)	1.20 (5)	1.14 (7)	1.11 (9)
	2	21,080	3.00 (5)	1.40 (5)	1.17 (6)	1.12 (8)
	3	83,760	3.83 (6)	2.00 (5)	1.17 (6)	1.14 (7)
	4	333,920	4.67 (6)	2.29 (7)	1.14 (7)	1.17 (6)
	5	1,333,440	5.17 (6)	2.57 (7)	1.50 (8)	1.17 (6)

results were obtained on other test cases and we adopt the Picard iteration henceforth.

To see the efficiency of the Schur complement approximation (2.27) we used in Section 2.3.2, we give the number of Krylov iterations for ALLU in Table 4.2. It can be observed that as γ increases, the preconditioner becomes a better approximation to the real Jacobian inverse and the preconditioner is mesh-independent.

Remark 4.1. It can be noted from Table 4.2 that ALLU seems to give a rather reasonable solver for $\gamma = 0$ (and thus with no penalisation of the unit-length constraint). One may wonder whether the example illustrated in this subsection is a good one for testing the application of augmented Lagrangian methods. Indeed, this is a simpler case but with a known exact

Table 4.2. ALLU: The average number of FGMRES iterations per nonlinear iteration for a nematic LC problem in a square slab using $[\mathbb{P}_2]^3$–$\mathbb{P}_1$ discretisation. Note here the last four columns are excerpted from Table 4.1 using the Picard iteration.

#refs	#dofs	γ							
		0	1	10	10^2	10^3	10^4	10^5	10^6
1	5,340	10.40	9.20	8.00	5.40	2.00	1.20	1.14	1.11
2	21,080	14.20	13.20	9.20	5.80	3.00	1.40	1.17	1.12
3	83,760	4.75	4.75	6.75	6.40	3.83	2.00	1.17	1.14
4	333,920	5.50	4.50	7.25	7.20	4.67	2.29	1.14	1.17
5	1,333,440	5.25	3.75	5.75	7.00	5.17	2.57	1.50	1.17

Table 4.3. ALMG-STAR: The average number of FGMRES iterations per nonlinear iteration (total Newton iterations) for the nematic LC problem in a square slab.

#refs	#dofs	γ			
		10^3	10^4	10^5	10^6
1	5,340	2.60 (5)	2.40 (5)	2.29 (7)	2.29 (7)
2	21,080	4.20 (5)	2.20 (5)	2.50 (6)	3.29 (7)
3	83,760	8.00 (5)	3.00 (5)	2.33 (6)	3.33 (6)
4	333,920	11.60 (5)	5.17 (6)	2.17 (6)	2.29 (7)
5	1,333,440	15.20 (5)	8.43 (7)	3.14 (7)	1.78 (9)

solution and it is intended for showing the efficiency and convergence rate of our proposed AL preconditioner. More complicated cases will be given later.

The performance of ALMG-STAR (utilising the augmented Lagrangian preconditioner with star patch as the relaxation in the multigrid algorithm) and ALMG-PBJ (utilising the augmented Lagrangian preconditioner with the point-block Jacobian relaxation in the multigrid algorithm) are illustrated in Tables 4.3 and 4.4, respectively, where both mesh independence for $\gamma = 10^6$ and γ-robustness are observed.

We also test the robustness of ALMG-STAR and ALMG-PBJ on other problem parameters, e.g., the twist elastic constant $K_2 > 0$ and the cholesteric pitch q_0. To this end, we continue $K_2 \in [0.2, 8]$ and $q_0 \in [0, 8]$ with step 0.1. We fix $\gamma = 10^6$, since it gives the best performance in Tables 4.3 and 4.4. The numerical results of ALMG-STAR and ALMG-PBJ in K_2- and q_0-continuation are shown in Figures 4.2 and 4.3, respectively.

Table 4.4. ALMG-PBJ: The average number of FGMRES iterations per nonlinear iteration (total Newton iterations) for the nematic LC problem in a square slab.

			γ		
#refs	#dofs	10^3	10^4	10^5	10^6
1	5,340	3.20 (5)	2.60 (5)	3.00 (6)	3.57 (7)
2	21,080	5.60 (5)	2.60 (5)	2.83 (6)	3.71 (7)
3	83,760	10.00 (5)	3.80 (5)	2.80 (5)	3.00 (6)
4	333,920	15.40 (5)	7.00 (5)	2.50 (6)	2.83 (6)
5	1,333,440	>100	11.83 (6)	5.00 (5)	2.83 (6)

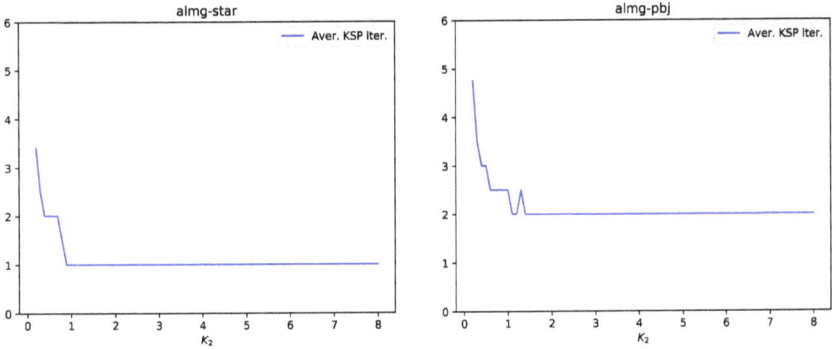

Fig. 4.2. Average number of FGMRES iterations per nonlinear iteration when continuing in K_2 for the LC problem in a square slab.

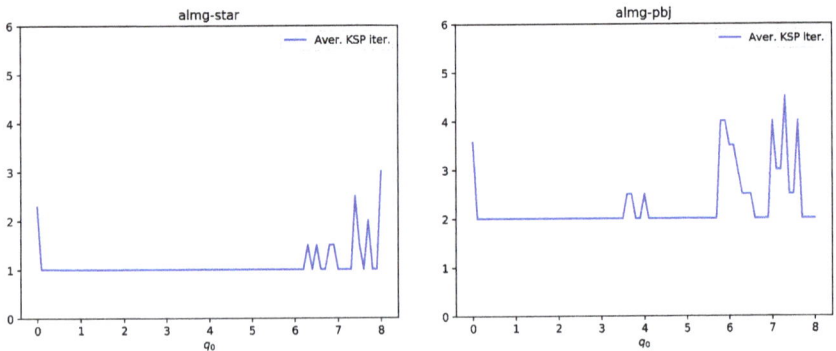

Fig. 4.3. Average number of FGMRES iterations per nonlinear iteration when continuing in q_0 for the LC problem in a square slab.

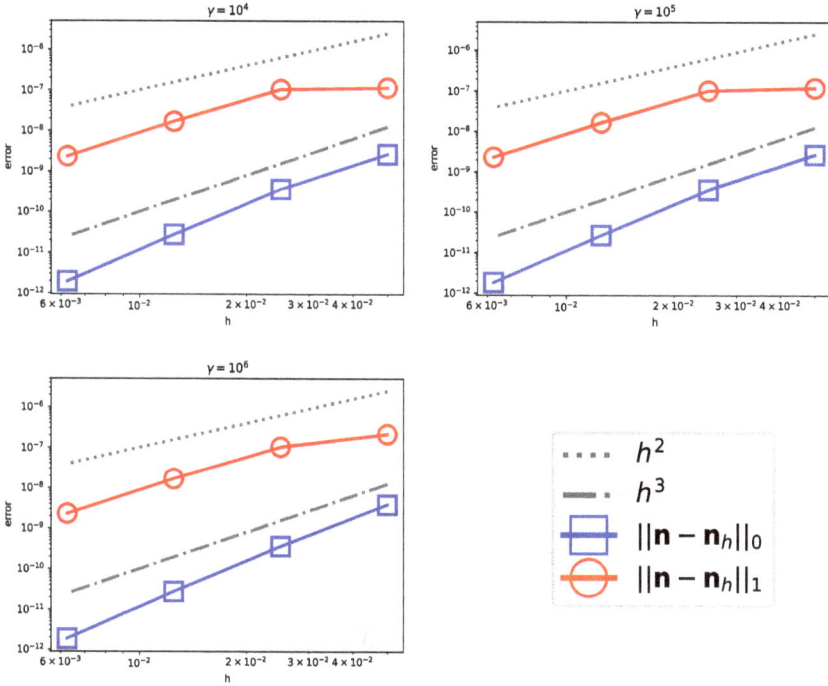

Fig. 4.4. The convergence of the computed director as the mesh is refined for the nematic LC problem in a square slab.

Clearly, a stable number of linear iterations is shown for both continuation experiments.

To examine the convergence order of the discretisation as a function of γ, we apply the ALMG-PBJ solver for $\gamma = 10^4, 10^5$ and 10^6. Note that the convergence result does not rely on the solver used. Figure 4.4 shows the L^2- and H^1-error between the computed director and the known analytical solution. We observe third-order convergence of the director in the L^2 norm and second-order convergence in the H^1 norm for all values of γ considered.

To investigate the computational efficiency of the AL approach, we compare our proposed AL-based solvers (ALMG-PBJ and ALMG-STAR) with a monolithic multigrid preconditioner using Vanka relaxation (Adler *et al.*, 2015a; Vanka, 1986) on each level (denoted as MGVANKA) in Table 4.5. Essentially, MGVANKA applies multigrid to the coupled director–multiplier problem, with an additive Schwarz relaxation organised around gathering all director dofs coupled to a given multiplier dof. All results are computed in serial. In our experiments, these two AL-based

Table 4.5. The computing time of ALMG-PBJ, ALMG-STAR and MGVANKA as a function of mesh refinement for the nematic LC problem in a square slab.

	Computing time (in minutes)					
#refs	1	2	3	4	5	6
#dofs	5,340	21,080	83,760	333,920	1,333,440	5,329,280
ALMG-PBJ	0.02	0.04	0.09	0.32	1.17	5.53
ALMG-STAR	0.02	0.07	0.23	0.79	2.95	12.86
MGVANKA	0.04	0.15	0.38	1.44	5.91	25.09

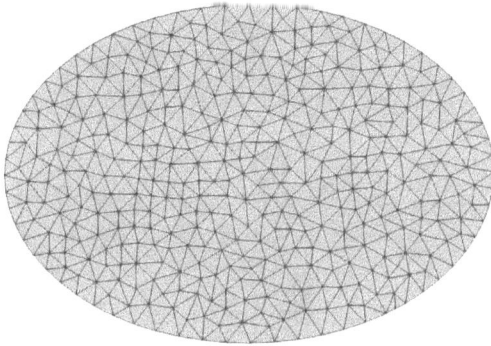

Fig. 4.5. The coarse mesh of the ellipse.

solvers outperform MGVANKA even for small problems of about five thousand dofs. In particular, ALMG-PBJ is the fastest method considered and is approximately five times faster than MGVANKA for a problem with about five million dofs. We also note that ALMG-STAR is slower than ALMG-PBJ, which is caused by the size of the star patch being larger than that of the point-block patch, requiring more work in the multigrid relaxation.

4.2.2 *Equal-constant nematic case in an ellipse*

Consider an ellipse of aspect ratio $3/2$ with strong anchoring boundary condition $\mathbf{n} = [0, 0, 1]^\top$ imposed on the entire boundary. We consider the equal-constant nematic case $K_1 = K_2 = K_3 = 1$, $q_0 = 0$ in the minimisation problem (2.1) to verify the theoretical results presented in the previous sections with corresponding discretisations. We use the initial guess $\mathbf{n}_0 = [0, 0, 0.8]^\top$ in the nonlinear iteration. The coarsest triangulation, generated in Gmsh (Geuzaine and Remacle, 2009), is illustrated in Figure 4.5.

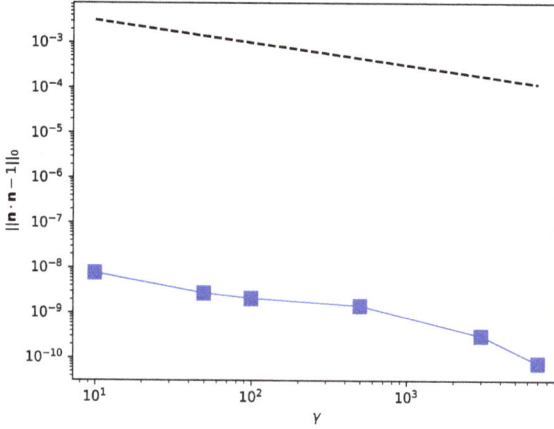

Fig. 4.6. Comparison of the computed constraint $\|\mathbf{n} \cdot \mathbf{n} - 1\|_0$ and the reference line $\mathcal{O}(\gamma^{-1/2})$ using the $[\mathbb{P}_1]^3$–$\mathbb{P}_1$ finite element pair for equal-constant nematic LC problems in an ellipse.

Table 4.6. ALLU: The average number of FGMRES iterations per nonlinear iteration for an equal-constant nematic problem in an ellipse using $[\mathbb{P}_2]^3$–$\mathbb{P}_1$ discretisation.

		γ							
#refs	#dofs	0	1	10	10^2	10^3	10^4	10^5	10^6
1	19,933	29.20	25.60	16.40	5.20	2.60	1.60	1.33	1.14
2	78,810	32.50	26.00	14.00	6.80	3.40	1.80	1.33	1.17
3	313,408	12.50	15.50	16.25	7.60	4.20	2.20	1.33	1.17
4	1,249,980	11.00	12.25	14.75	8.40	4.80	2.60	1.40	1.17
5	4,992,628	12.33	13.33	11.75	8.00	5.20	3.00	1.50	1.14

To verify our theoretical results about the improvement of the discrete enforcement of the constraint in Section 2.3.3, we vary the penalty parameter γ, use one refinement for the fine mesh, and employ the $[\mathbb{P}_1]^3$–$\mathbb{P}_1$ element. The data is plotted in Figure 4.6. The L^2-norm $\|\mathbf{n} \cdot \mathbf{n} - 1\|_0$ of the residual of the constraint decreases as γ grows and scales like $\mathcal{O}(\gamma^{-1/2})$ as expected.

The efficiency of the Schur complement approximation of Section 2.3.2 for the $[\mathbb{P}_2]^3$–$\mathbb{P}_1$ element can be observed in Table 4.6.

Tables 4.7 and 4.8 demonstrate the robustness of ALMG-STAR and ALMG-PBJ with respect to γ and mesh refinement for the $[\mathbb{P}_2]^3$–$\mathbb{P}_1$ element. It can be seen that both solvers are robust with respect to the penalty parameter γ and with respect to the mesh size h for $\gamma = 10^6$. The number

Table 4.7. ALMG-STAR: The average number of FGMRES iterations per nonlinear iteration (total nonlinear iterations) for equal-constant nematic problem in an ellipse using $[\mathbb{P}_2]^3$–$\mathbb{P}_1$ discretisation.

#refs	#dofs	γ			
		10^3	10^4	10^5	10^6
1	19,933	2.60 (5)	1.60 (5)	1.80 (5)	1.67 (6)
2	78,810	4.40 (5)	1.80 (5)	1.60 (5)	1.50 (6)
3	313,408	6.80 (5)	3.20 (5)	1.50 (6)	1.50 (6)
4	1,249,980	10.00 (5)	4.67 (6)	1.80 (5)	1.50 (6)
5	4,992,628	14.40 (5)	7.50 (6)	4.20 (5)	1.33 (6)

Table 4.8. ALMG-PBJ: The average number of FGMRES iterations per Newton iteration (total Newton iterations) for equal-constant nematic problem in an ellipse using $[\mathbb{P}_2]^3$–$\mathbb{P}_1$ discretisation.

#refs	#dofs	γ			
		10^3	10^4	10^5	10^6
1	19,933	3.80 (5)	2.60 (5)	2.60 (5)	2.80 (5)
2	78,810	6.80 (5)	3.20 (5)	2.60 (5)	2.60 (5)
3	313,408	9.00 (5)	5.00 (5)	2.60 (5)	2.60 (5)
4	1,249,980	14.80 (5)	8.20 (5)	3.80 (5)	2.40 (5)
5	4,992,628	19.00 (5)	11.60 (5)	6.80 (5)	2.50 (6)

of nonlinear iterations and the number of FGMRES iterations per nonlinear step remain stable.

4.3   Summary

In this chapter, we presented numerical results of our proposed AL preconditioner for two examples of LC problems in two dimensions (an ellipse and a square slab). We demonstrated the effectiveness and robustness (regarding problem-related parameters, the elastic constant K_2 and the cholesteric pitch q_0, and the mesh size h) of the preconditioner. We also tested the efficiency of preconditioners with star and point block patches and gave the numerical verification of the improvement of the constraint proven in Section 2.3.3.

This part (from Chapters 2–4) resolves the difficulty of solving a unit-length constrained minimisation problem of the Oseen–Frank model for LC by applying augmented Lagrangian methods. It provides a viable approach to construct efficient and robust solvers for liquid crystal problems involving Oseen–Frank models, although the complexity can rapidly increase when it comes to more sophisticated phases requiring the coupling with other order parameters, e.g., in ferronematics and smectics. In the remainder of this book, we consider another modelling theory which avoids the imposition of unit-length constraint for a vector field and instead turn to the so-called **Q**-tensor theory.

Part 2

Ferronematic Liquid Crystals

Chapter 5

A Mathematical Model of Ferronematics[*]

In the previous part, we considered the Oseen–Frank model for cholesteric and nematic liquid crystals. This model uses a vector-valued order parameter and only applies to uniaxial LC, i.e., where only one direction of molecular alignment is preferred. In fact, the Oseen–Frank formulation is known to be limited, in the sense that it can only account for point defects, but not the more complicated line or surface defects that are observed experimentally (Majumdar and Zarnescu, 2010). One can simply check this by observing that the Oseen–Frank free energy $\mathcal{J}^{OF}(\mathbf{n})$ (2.1), with equal Frank constants and zero cholesteric pitch q_0, blows up for the line defect $\mathbf{n} = [x, y, 0]^\top / \sqrt{x^2 + y^2}$, while the energy functional is well-defined for the point defect $\mathbf{n} = [x, y, z]^\top / \sqrt{x^2 + y^2 + z^2}$. Another potential drawback of this theory is its inability of representing half-charge defects, due to the presence of director discontinuities in these defects (Ball, 2017), which cannot be characterised by a continuous vector field. For example, around a $\pm 1/2$ defect where $\mathbf{n}$ rotates by $\pm\pi$ degrees, a discontinuity line (i.e., *branch cut* in Ball and Zarnescu (2008)) where $\mathbf{n}$ reverses sign must exist.

Hence, to better characterise the defect structure, particularly in more complex liquid crystal phases and applications, we instead use a more complete phenomenological description for LC: the Landau–de Gennes (LdG) theory (de Gennes, 1969, 1974), which can account for both uniaxial and biaxial (having more than one preferred direction of molecular alignment) phases. The LdG theory is widely used in the modelling of phase transitions in liquid crystals (Biscari *et al.*, 2007; de Gennes, 1969) and we thus adopt it for ferronematics and smectics in the remainder of this book.

---

[*]This part of work majorly expands on the author's work in Dalby *et al.* (2022).

In this part, we consider the case of ferronematics and we first give an introduction on some details of the LdG theory to prepare ourselves for modelling ferronematics.

5.1   The Landau–de Gennes Model

In this framework, the state of nematic LC is modelled by a symmetric, traceless tensor field $\mathbf{Q} : \Omega \to S_0$, known as the tensor order parameter. Here, S_0 denotes the set of all symmetric, traceless $d \times d$ matrices. We consider a three-dimensional domain $\Omega \subset \mathbb{R}^3$ (i.e., $d = 3$) filled with liquid crystal as an example in this subsection, the two dimensional case is analogous. The eigenvectors $\mathbf{e}_1, \mathbf{e}_2$ and $\mathbf{e}_3$ of $\mathbf{Q} \in S_0$ are the directions of the preferred molecular orientations and their associated eigenvalues λ_1, λ_2 and λ_3 represent the degree of order along each corresponding direction (Mottram and Newton, 2014).

We say that liquid crystals are (a) *isotropic* if $\mathbf{Q}$ has three equal eigenvalues, i.e., $\lambda_1 = \lambda_2 = \lambda_3$ (and hence, $\mathbf{Q} = 0$). They are (b) *uniaxial* when $\mathbf{Q}$ has two equal nonzero eigenvalues (say, $2|\lambda_1| = 2|\lambda_2| = |\lambda_3|$, thus λ_3 is the major eigenvalue). Such uniaxial $\mathbf{Q}$-tensors can be written in the special form

$$\mathbf{Q} = s \left(\mathbf{n} \otimes \mathbf{n} - \frac{\mathbf{I}_3}{3} \right), \quad s : \Omega \to \mathbb{R}, \quad \mathbf{n} : \Omega \to \mathcal{S}^2,$$

where $s = \frac{3}{2}\lambda_3$. Finally, they are (c) *biaxial* when $\mathbf{Q}$ has three distinct eigenvalues. A biaxial $\mathbf{Q}$-tensor can always be represented by

$$\mathbf{Q} = s \left(\mathbf{n} \otimes \mathbf{n} - \frac{1}{3}\mathbf{I}_3 \right) + t \left(\mathbf{r} \otimes \mathbf{r} - \frac{1}{3}\mathbf{I}_3 \right), s, t : \Omega \to \mathbb{R}, \mathbf{n}, \mathbf{r} : \Omega \to \mathcal{S}^2.$$
$$(5.1)$$

The LdG energy for nematic LC is of the form

$$\mathcal{J}^{LdG}(\mathbf{Q}) = \int_\Omega \left\{ f_n^e(\nabla \mathbf{Q}) + f_n^b(\mathbf{Q}) \right\},$$

where f_n^e and f_n^b correspond to the nematic *elastic* and *bulk* energy densities to be defined in the following. Note that the minimisation problem with functional $\mathcal{J}^{LdG}$ is unconstrained, as opposed to the constrained minimisation problem (2.1) in the Oseen–Frank theory.

The elastic part consists of three independent quadratic terms with respect to the first partial derivatives of components of $\mathbf{Q}$. Specifically, we take the form

$$f_n^e(\nabla \mathbf{Q}) = \frac{1}{2}\{K_a \mathbf{Q}_{ij,n}\mathbf{Q}_{ij,n} + K_b \mathbf{Q}_{ij,j}\mathbf{Q}_{in,n} + K_c \mathbf{Q}_{ij,n}\mathbf{Q}_{in,j}\}, \qquad (5.2)$$

where K_a, K_b and K_c are *elastic* constants depending on the material. Here, we adopted the Einstein summation convention for repeated indices.

The bulk energy density is typically a truncated expansion in the scalar invariants of $\mathbf{Q}$ and accounts for bulk effects. One commonly used form (de Gennes, 1974; Mottram and Newton, 2014) is

$$f_n^b(\mathbf{Q}) = \frac{l_a}{2}\operatorname{tr}\left(\mathbf{Q}^2\right) - \frac{l_b}{3}\operatorname{tr}\left(\mathbf{Q}^3\right) + \frac{l_c}{4}\left(\operatorname{tr}\left(\mathbf{Q}^2\right)\right)^2. \qquad (5.3)$$

Here, $l_b, l_c > 0$ are material-dependent *bulk* constants, independent of temperature, whereas $l_a < 0$ depends on the temperature.

Taking $K_a = K_{LdG}$, $K_b = K_c = 0$ in (5.2), we obtain the *one-constant* form of the LdG energy for nematic LC:

$$\mathcal{J}^{LdG}(\mathbf{Q}) = \int_\Omega \left\{ \frac{K_{LdG}}{2}|\nabla \mathbf{Q}|^2 + \frac{l_a}{2}\operatorname{tr}\left(\mathbf{Q}^2\right) - \frac{l_b}{3}\operatorname{tr}\left(\mathbf{Q}^3\right) + \frac{l_c}{4}\left(\operatorname{tr}\left(\mathbf{Q}^2\right)\right)^2 \right\}, \qquad (5.4)$$

which will be employed in several places, e.g., (5.6) in ferronematics and our new proposed smectic model (8.10).

5.2 Full Model of Ferronematics

To start with the first application of the LdG theory in this book, we now briefly introduce ferronematic materials and their modelling.

Nematic LC are anisotropic materials that can respond to applied external fields and are thus suitable for a wide range of electro-optic devices, especially liquid crystal displays. One immediate example is the twisted nematic display (Dunmur and Sluckin, 2011, Technical Box 10.1) where the display is switched on and off by activating or deactivating an electric field applied to the nematic LC. In fact, this response relies on the dielectric anisotropy of nematics, that is to say, the directional response to external electric fields (Lagerwall and Scalia, 2012). In contrast, when exposed to magnetic fields, their responses are much weaker (perhaps seven orders of magnitude smaller) than that of electric fields (Stewart, 2004). Consequently, nemato-magnetic coupling effect has not been extensively exploited for nematic applications, e.g., sensors, displays, microfluidics etc.

One pioneering work dating back to 1970 by Brochard and de Gennes (1970) found that a suspension of magnetic nanoparticles (MNPs) in a nematic phase can induce a spontaneous magnetisation in the absence of an external magnetic field, and substantially enhance the nemato-magnetic response. They referred to this new class of materials as *ferronematics*, possessing the useful feature that the nematic and magnetic order parameters are strongly coupled. Subsequently, there were some notable theoretical contributions regarding ferronematic modelling made by Burylov and Raikher (1995) and Calderer *et al.* (2014), where continuum models were discussed and analysed. Meanwhile, some experiments about ferronematics were also realised by Rault *et al.* (1970), and more recently by Mertelj *et al.* (2013). Due to their special responses in the absence of any external magnetic fields, ferronematics may find potential use in magneto-optic devices.

In this chapter, we study a dilute suspension of MNPs in a three-dimensional nematic-filled channel, $\tilde{\Omega} = [-L, L] \times [-D, D] \times [0, G]$, where $L \gg D$ is the length of the channel, D is the width and G is the height. Since $L \gg D$, it is reasonable to assume that molecules are uniform along the length and across the height of the channel, and there are no boundary constraints imposed at the two ends $x = \pm L$. Thus, we can restrict ourselves to a one-dimensional geometry: $\bar{\Omega} = [-D, D]$. We then rescale this domain to $\Omega = [-1, 1]$ for simplicity, similarly to Bisht *et al.* (2019).

The suspended MNPs generate a spontaneous magnetisation (in the absence of any external magnetic fields) by means of the nemato-magnetic coupling. In this system, there are two order parameters: (i) a nematic tensor parameter $\mathbf{Q} : \Omega \to S_0$ (symmetric, traceless 2×2 matrices), indicating the preferred molecular alignment of the director in the nematic host and (ii) a vector-valued magnetic order parameter $\mathbf{M} : \Omega \to \mathbb{R}^2$, $\mathbf{M} = (M_1, M_2)^\top$, generated by the suspended MNPs.

In the uniaxial case, as discussed above the nematic order parameter $\mathbf{Q}$ can be written as

$$\mathbf{Q} = s(2\mathbf{n} \otimes \mathbf{n} - \mathbf{I}_2), \qquad (5.5)$$

where s is a scalar order parameter and $\mathbf{n}$ is the nematic director. Here, s can be interpreted as a measure of the degree of the orientational order for director $\mathbf{n}$, so that the nodal set of s (i.e., where $s = 0$) indicates the presence of nematic defects (where an orientation is not well-defined). We denote the two independent components of $\mathbf{Q}$ by Q_{11} and Q_{12} such that

$$Q_{11} = s \cos 2\vartheta, \quad Q_{12} = s \sin 2\vartheta,$$

where $\mathbf{n} = (\cos\vartheta, \sin\vartheta)$ and ϑ denotes the angle between $\mathbf{n}$ and the horizontal axis. To avoid writing $\mathbf{Q}$ in the matrix form $\left[\begin{smallmatrix} Q_{11} & Q_{12} \\ Q_{12} & -Q_{11} \end{smallmatrix}\right]$, we henceforth label $\mathbf{Q}$ in terms of its two independent components (Q_{11}, Q_{12}), when this causes no confusions. Consequently, we use the vector norm $|\mathbf{Q}| = \sqrt{Q_{11}^2 + Q_{12}^2}$, as opposed to the usual matrix norm. The conventional definition of the vector norm is adopted for the magnetisation vector $\mathbf{M}$, that is to say, $|\mathbf{M}| = \sqrt{M_1^2 + M_2^2}$.

By following the methods in Bisht *et al.* (2019) and Mertelj *et al.* (2013), we use the total rescaled and dimensionless ferronematic energy of the form

$$
\begin{aligned}
\mathcal{J}^{fer}(Q_{11}, Q_{12}, M_1, M_2) := \int_\Omega \Bigg\{ &\frac{k_1}{2}\left[\left(\frac{dQ_{11}}{dy}\right)^2 + \left(\frac{dQ_{12}}{dy}\right)^2\right] \\
&+ \left(Q_{11}^2 + Q_{12}^2 - 1\right)^2 \\
&+ \frac{\xi k_2}{2}\left[\left(\frac{dM_1}{dy}\right)^2 + \left(\frac{dM_2}{dy}\right)^2\right] \\
&+ \frac{\xi}{4}\left(M_1^2 + M_2^2 - 1\right)^2 \\
&- cQ_{11}\left(M_1^2 - M_2^2\right) - 2cQ_{12}M_1M_2 \Bigg\}\, dy,
\end{aligned}
$$

$$(5.6)$$

and the associated minimisation problem is

$$\min \mathcal{J}^{\text{fer}}(Q_{11}, Q_{12}, M_1, M_2). \tag{5.7}$$

Here, $k_1 > 0$ and $k_2 > 0$ are scaled elastic constants (in practice, $k_1 > k_2$ since the nematic effect dominates in ferronematics), $\xi > 0$ is a parameter that weighs the relative strength of the nematic and magnetic energies, and c is a coupling parameter. Since we consider a dilute suspension of MNPs, there are only "small" interactions between MNPs while the nemato-magnetic interactions are taken into account through the coupling energy term. Therefore, we can see that the magnetic energy part is not dominating and it is reasonable that $\xi \leq 1$ (Calderer *et al.*, 2014).

The ferronematic free energy is a sum of three energetic contributions: a LdG-type nematic energy of $\mathbf{Q}$, a magnetisation energy of $\mathbf{M}$, and a coupling energy between $\mathbf{Q}$ and $\mathbf{M}$. Substituting the uniaxial expression

(5.5) into the coupling energy, we observe that

$$-cQ_{11}\left(M_1^2 - M_2^2\right) - 2cQ_{12}M_1M_2 \propto -c\left(\mathbf{n}\cdot\mathbf{M}\right)^2.$$

In this part of work, we only focus on positive coupling $(c > 0)$ so that the coupling energy favours co-alignment between the nematic director $\mathbf{n}$ and magnetic vector $\mathbf{M}$.

Furthermore, we consider imposing Dirichlet boundary conditions for both $\mathbf{Q}$ and $\mathbf{M}$ on the ends $y = \pm 1$:

$$Q_{11}\left(-1\right) - M_1\left(-1\right) = 1, \tag{5.8a}$$

$$Q_{12}(-1) = Q_{12}(1) = M_2(-1) = M_2(1) = 0, \tag{5.8b}$$

$$Q_{11}\left(1\right) = M_1\left(1\right) = -1. \tag{5.8c}$$

Here, the boundary conditions for $\mathbf{Q}$ correspond to $\mathbf{n} = (1,0)$ on $y = -1$ and $\mathbf{n} = (0,1)$ on $y = 1$, that is to say, we are essentially enforcing planar boundary conditions for $\mathbf{Q}$ at $y = -1$ and homeotropic boundary conditions at the other end $y = +1$. Meanwhile, the boundary conditions for $\mathbf{M}$ describe a π-rotation of magnetic orientation between the bounding plates $y = \pm 1$. Then, the admissible space of the minimisation problem (5.7) is given by

$$\mathcal{A}_f = \big\{\mathbf{Q} \in H^1\left(\Omega, S_0\right), \mathbf{M} \in H^1\left(\Omega, \mathbb{R}^2\right),$$

$$\mathbf{Q} \text{ and } \mathbf{M} \text{ satisfy the boundary conditions (5.8)}\big\}. \tag{5.9}$$

We are interested in the local or global energy minimisers $(\mathbf{Q}, \mathbf{M})$, being stable and potentially observable, of the ferronematic free energy (5.6) in the admissible space $\mathcal{A}_f$. In fact, they are *classical* solutions (which can be verified by elliptic regularity, suitable Sobolev embeddings and bootstrapping arguments) of the associated Euler–Lagrange equations

$$k_1\frac{\mathrm{d}^2 Q_{11}}{\mathrm{d}y^2} = 4Q_{11}(Q_{11}^2 + Q_{12}^2 - 1) - c\left(M_1^2 - M_2^2\right), \tag{5.10a}$$

$$k_1\frac{\mathrm{d}^2 Q_{12}}{\mathrm{d}y^2} = 4Q_{12}(Q_{11}^2 + Q_{12}^2 - 1) - 2cM_1M_2, \tag{5.10b}$$

$$\xi k_2\frac{\mathrm{d}^2 M_1}{\mathrm{d}y^2} = \xi M_1\left(M_1^2 + M_2^2 - 1\right) - 2cQ_{11}M_1 - 2cQ_{12}M_2, \tag{5.10c}$$

$$\xi k_2\frac{\mathrm{d}^2 M_2}{\mathrm{d}y^2} = \xi M_2\left(M_1^2 + M_2^2 - 1\right) + 2cQ_{11}M_2 - 2cQ_{12}M_1. \tag{5.10d}$$

Remark 5.1. For simplicity and brevity, we take $k_1 = k_2 = k$ and $\xi = 1$ hereafter. One can tackle the cases of $k_1 \neq k_2$ and $\xi \neq 1$ using similar mathematical methods.

An immediate question arises regarding the existence and uniqueness of minimisers of the problem (5.7) in the admissible space $\mathcal{A}_f$. The existence result is proven in [Dalby *et al.* (2022)] via the direct method of the calculus of variations. Uniqueness holds for sufficiently large k. We quote the theorem below for self-containment.

Theorem 5.1 (Dalby *et al.*, 2022; Uniqueness of minimisers for sufficiently large k). *For a fixed c and for k sufficiently large, there exists a unique critical point (and hence global minimiser) of the ferronematic free energy (5.6) in the admissible space (5.9).*

Moreover, a maximum principle for the solutions $(Q_{11}, Q_{12}, M_1, M_2)$ of the system (5.10a)–(5.10d) is obtained in Dalby *et al.* (2022) and we include this result in the following so that we can numerically verify it later in Chapter 7.

Theorem 5.2 (Dalby *et al.*, 2022; Maximum principle). *There exists an L^∞ bound for the solutions $(Q_{11}, Q_{12}, M_1, M_2)$ of the system (5.10a)–(5.10d) subject to the boundary conditions (5.8). Specifically,*

$$Q_{11}^2(y) + Q_{12}^2(y) \leq (\rho^*)^2, \quad M_1^2(y) + M_2^2(y) \leq 1 + 2c\rho^* \quad \forall y \in [-1, 1],$$
(5.11)

where ρ^* is given by

$$\rho^* = \left(\frac{c}{8} + \sqrt{\frac{c^2}{64} - \frac{1}{27}\left(1 + \frac{c^2}{2}\right)^3} \right)^{\frac{1}{3}} + \left(\frac{c}{8} - \sqrt{\frac{c^2}{64} - \frac{1}{27}\left(1 + \frac{c^2}{2}\right)^3} \right)^{\frac{1}{3}}.$$
(5.12)

Remark 5.2. We will verify the L^∞ bound (5.11) numerically for each solution in Chapter 7.

With the uniqueness and maximum principle results at hand, we can notice that in the $k \to \infty$ limit, it is theoretically expected to have only one minimiser of the ferronematic free energy (5.6) and there is a k-independent L^∞ bound (given by (5.11)) for $\mathbf{Q}, \mathbf{M}$. Moreover, in this limit, one can easily see that the Euler–Lagrange equations (5.10a)–(5.10d) reduce to the Laplace equations

$$\frac{\mathrm{d}^2 Q_{11}}{\mathrm{d}y^2} = 0, \quad \frac{\mathrm{d}^2 Q_{12}}{\mathrm{d}y^2} = 0,$$
$$\frac{\mathrm{d}^2 M_1}{\mathrm{d}y^2} = 0, \quad \frac{\mathrm{d}^2 M_2}{\mathrm{d}y^2} = 0,$$

(5.13)

subject to the boundary conditions (5.8). This Laplace system then admits a unique solution:

$$(\mathbf{Q}^\infty, \mathbf{M}^\infty) = (Q_{11}^\infty, Q_{12}^\infty, M_1^\infty, M_2^\infty) = (-y, 0, -y, 0), \tag{5.14}$$

where Q_{12}, M_2 are zero-valued and Q_{11}, M_1 are linear profiles. The solution (5.14) is also referred to as an *order reconstruction* solution, with only two degrees of freedom (Q_{11}, M_1) reduced from the full four degrees of freedom $(Q_{11}, Q_{12}, M_1, M_2)$. We will discuss this reduced system further in Section 5.3. The convergence result regarding the limit regime $k \to \infty$ is proven in Dalby *et al.* (2022) using the method of sub- and super-solutions and we quote the theorem below, which is to be numerically validated as well in Chapter 7.

Theorem 5.3 (Dalby *et al.*, 2022; Convergence result of $k \to \infty$). *Assume k is sufficiently large so that the uniqueness result Theorem 5.1 holds. Let $(\mathbf{Q}^k, \mathbf{M}^k)$ be the unique solution of the Euler–Lagrange equations (5.10a)–(5.10d) in the admissible space (5.9), subject to the boundary conditions (5.8). Then $(\mathbf{Q}^k, \mathbf{M}^k)$ converge to $(\mathbf{Q}^\infty, \mathbf{M}^\infty)$ as $k \to \infty$ with the following estimates:*

$$\forall j = 1, 2, \quad \|Q_{1j}^k - Q_{1j}^\infty\|_\infty \le \alpha_1 k^{-1}, \ \|M_j^k - M_j^\infty\|_\infty \le \alpha_2 l^{-1},$$

for positive constants α_1, α_2 independent of k.

Remark 5.3. It implies that when k is sufficiently large, there is only one unique minimiser of the form (5.14) which gives a linear order reconstruction profile.

The case of $k \to 0$ is more complicated due to the non-uniqueness of solutions and in fact its convergence information requires more delicate Γ-convergence analysis. However, some preliminary properties about the limiting profile for $k \to 0$ can be obtained by examining the so-called *bulk minimisers* that minimise the bulk energy (i.e., eliminating all gradient terms in the ferroenematic full energy (5.6)):

$$F_b(Q_{11}, Q_{12}, M_1, M_2) := \left(Q_{11}^2 + Q_{12}^2 - 1\right)^2 + \frac{1}{4}\left(M_1^2 + M_2^2 - 1\right)^2$$
$$- cQ_{11}\left(M_1^2 - M_2^2\right) - 2cQ_{12}M_1M_2. \qquad (5.15)$$

Substituting the parametrisation

$$Q_{11} = \rho\cos(\theta), Q_{12} = \rho\sin(\theta),$$
$$M_1 = \sigma\cos(\phi), M_2 = \sigma\sin(\phi), \qquad (5.16)$$

into (5.15), we can deduce that the minimisers of F_b belong to the set

$$\mathcal{M}_{\min} := \{(Q_{11}, Q_{12}, M_1, M_2) = (\rho^*\cos(\theta), \rho^*\sin(\theta),$$
$$\sqrt{1 + 2c\rho^*}\cos(\phi), \sqrt{1 + 2c\rho^*}\sin(\phi)) :$$
$$\theta = 2\phi + 2z\pi, \text{ for } z \in \mathbb{Z}\},$$

where ρ^* is given by (5.12). Thus, we can define the limiting minimisers for $k \to 0$ as

$$\mathbf{Q}^f(c, y) = \rho^*(\cos(2\phi(y)), \sin(2\phi(y))),$$
$$\mathbf{M}^f(c, y) = \sqrt{1 + 2c\rho^*}(\cos(\phi(y)), \sin(\phi(y))), \qquad (5.17)$$

where there are two choices of ϕ due to the imposed boundary conditions for $\mathbf{M}^f$:

$$\frac{d^2\phi}{dy^2} = 0, \qquad (5.18a)$$

$$\phi(-1) = 0, \phi(1) = \pi \quad \text{or} \quad \phi(-1) = 0, \phi(1) = -\pi, \qquad (5.18b)$$

$$\theta - 2\phi = 2z\pi. \qquad (5.18c)$$

Remark 5.4. It is obvious from the definition (5.17) of the limiting minimisers for $k \to 0$ that neither $\mathbf{Q}$ nor $\mathbf{M}$ vanishes since ρ^* is nonzero.

Therefore, we expect that the energy minimisers $(\mathbf{Q}^f, \mathbf{M}^f)$ of the full energy (5.6) should converge to one of the defined limiting minimisers in (5.17) almost everywhere as $k \to 0$. The exception happens close to the boundary end points $y = \pm 1$ (due to the incompatible boundary conditions with the limiting minimisers) or at interior points that are associated with jumps in $(2\phi - \theta)$ (since $(2\phi - \theta)$ is only constrained to be an even multiple of 2π in the $k \to 0$ limit). The numerical verification of this hypothesis is illustrated in Section 7.2.

5.3   Reduced Model: Order Reconstruction

The previous section concerns the full ferronematic problem (5.10a)–(5.10d) with four degrees of freedom $(\mathbf{Q}, \mathbf{M}) = (Q_{11}, Q_{12}, M_1, M_2)$, i.e., four scalar unknowns. One can observe that profiles with $Q_{12} = M_2 = 0$ can always contribute to a branch of solutions of the Euler–Lagrange equations (5.10a)–(5.10d). We refer to these solutions with only two degrees of freedom, $(\mathbf{Q}, \mathbf{M}) = (Q_{11}, 0, M_1, 0)$ as *order reconstruction* (OR) solutions. This leads to the following reduced functional, denoted as the *OR* energy, from the full energy (5.6):

$$E(Q_{11}, M_1) := \int_{-1}^{1} \left\{ \frac{k}{2} \left(\frac{dQ_{11}}{dy} \right)^2 + \frac{k}{2} \left(\frac{dM_1}{dy} \right)^2 + (Q_{11}^2 - 1)^2 \right.$$
$$\left. + \frac{1}{4} \left(M_1^2 - 1 \right)^2 - cQ_{11}M_1^2 \right\} dy, \tag{5.19}$$

subject to the boundary conditions

$$Q_{11}(-1) = M_1(-1) = 1,$$
$$Q_{11}(1) = M_1(1) = -1, \tag{5.20}$$

in the admissible space

$$\mathcal{A}'_f = \{Q_{11}, M_1 \in H^1(\Omega, \mathbb{R}),$$
$$Q_{11} \text{ and } M_1 \text{ satisfy the boundary conditions (5.20)}\}. \tag{5.21}$$

Consequently, OR solutions are classical solutions of the following coupled ordinary differential equations,

$$k_1 \frac{d^2 Q_{11}}{dy^2} = 4Q_{11}(Q_{11}^2 - 1) - cM_1^2,$$
$$k_2 \frac{d^2 M_1}{dy^2} = M_1(M_1^2 - 1) - 2cQ_{11}M_1. \tag{5.22}$$

Remark 5.5. The reason why we are interested in the OR solutions is not only due to a reduction of unknowns that benefits our subsequent analysis, but also due to one of their special solutions, the so-called *domain wall* (i.e., $\mathbf{Q} = \mathbf{M} = \mathbf{0}$) profiles that separate *polydomains*, i.e., distinctly ordered domains. A nematic (respectively, magnetic) domain wall is a point $y = y^* \in (-1, 1)$ such that $\mathbf{Q}(y^*) = (Q_{11}(y^*), Q_{12}(y^*)) = 0$ (respectively, $\mathbf{M}(y^*) = 0$).

One can note from our applied inhomogeneous boundary conditions (5.20) for Q_{11} (resp. M_1) that there must exist an interior point, $y^* \in (-1, 1)$ such that $Q_{11}(y^*) = 0$ (resp. $M_1(y^*) = 0$) since $Q_{11}(-1) = M_1(-1) = 1$ and $Q_{11}(1) = M_1(1) = -1$. That is to say, we expect to see nematic and magnetic interior domain walls for the solutions $(\mathbf{Q}, \mathbf{M})$. Moreover, these domain walls can occur at different points (which we shall demonstrate in Chapter 7). In fact, using the parameterisation

$$Q_{11} = \rho \cos(\theta), Q_{12} = \rho \sin(\theta),$$
$$M_1 = \sigma \cos(\phi), M_2 = \sigma \sin(\phi),$$

(5.23)

we can notice that $Q_{12} = M_2 = 0$ implies $\theta = z_1 \pi$ and $\phi = z_2 \pi$ for some integers z_1, z_2. Furthermore, due to the imposed inhomogeneous boundary conditions, there is necessarily a domain wall in $\mathbf{Q}$ such that $\theta = 2z_1 \pi$ on one side of the domain wall containing the end point $y = -1$, and $\theta = (2z_2 + 1)\pi$ (for some integers z_1, z_2) on the other side of the domain wall containing the end point $y = 1$; analogously, there is a domain wall in $\mathbf{M}$ that separates two polydomains, with $\phi = 2z_1 \pi$ and $\phi = (2z_2 + 1)\pi$ for some integers z_1 and z_2 respectively.

Similarly, there are some qualitative results regarding the existence, uniqueness, maximum principle and instability of the OR solutions, proven in detail by Dalby and Majumdar (Dalby *et al.*, 2022). We again quote the following result so that we can numerically verify it in Chapter 7.

Theorem 5.4 (Dalby *et al.*, 2022; Uniqueness and maximum principle). *For sufficiently large k and a fixed positive c, the OR solution $(\mathbf{Q}^{\mathrm{OR}}, \mathbf{M}^{\mathrm{OR}}) := (Q_{11}^*, 0, M_1^*, 0)$ is the unique critical point, and hence, global minimiser of the energy (5.6), as in Theorem 5.1. Moreover, we have the L^∞ bound*

$$|Q_{11}(y)| \le \rho^*, \ M_1^2(y) \le 1 + 2c\rho^* \quad \forall y \in [-1, 1], \tag{5.24}$$

where ρ^* is given by (5.12).

It follows from Theorem 5.4 that the OR solution is the global minimiser for sufficiently large k and there is an L^∞ bound for Q_{11}, M_1. However, the OR solution loses its stability as k decreases, similarly to the study of the pure nematic case (i.e., $c = 0$) in Canevari *et al.* (2019) and Lamy (2014). We include the result below for self containment.

Theorem 5.5 (Dalby *et al.*, 2022; Instability of the OR solution). *For sufficiently small k and a fixed positive c, the OR energy minimiser, $(\mathbf{Q}^{OR}, \mathbf{M}^{OR})$, is an unstable critical point of the full energy (5.6), in the full admissible space (5.9).*

The convergence result for $k \to 0$ limiting regime is given by Dalby and Majumdar in Dalby *et al.* (2022) using Γ-convergence methods by directly following Wang *et al.* (2019, Proposition 4.1). More precisely, when k is very small, the minimisers to the OR energy (5.19) is closely related to the OR bulk minimisers:

$$\mathbf{p}^* = (Q_{11}, M_1) = \left(\rho^*, \sqrt{1 + 2c\rho^*}\right), \text{ or } \mathbf{p}^{**}(Q_{11}, M_1) = \left(\rho^*, -\sqrt{1 + 2c\rho^*}\right).$$
(5.25)

Remark 5.6. Note that these profiles in (5.25) are not compatible with the boundary conditions (5.20). Thus, there are necessarily boundary layers close to $y = \pm 1$ in the OR energy minimisers as $k \to 0$.

We do not include a detailed description of the convergence results as $k \to 0$, however, we can numerically demonstrate that the OR energy minimiser converges to $\mathbf{p}^*$ almost everywhere as it has the least transition costs. To see this, we need to define the non-negative OR bulk energy:

$$\tilde{f}(Q_{11}, M_1) := \left(Q_{11}^2 - 1\right)^2 + \frac{1}{4}\left(M_1^2 - 1\right)^2 - cQ_{11}M_1^2 - \beta(c) \geq 0,$$
(5.26)

where the c-dependent constant $\beta(c)$ is the minimum value of the OR bulk potential.

Following Braides (2006) and Wang *et al.* (2019), we let $\mathbf{p} = (Q_{11}, M_1)$ and define the following metric ω (which is in fact the geodesic distance associated with the Riemannian metric $\tilde{f}^{1/2}$ [Wang *et al.* (2019)]) in the p-plane, for any two points $\mathbf{p}_0, \mathbf{p}_1 \in \mathbb{R}^2$:

$$\omega\left(\mathbf{p}_0, \mathbf{p}_1\right) = \inf \left\{ \int_{-1}^{1} \tilde{f}^{1/2}\left(\mathbf{p}(t)\right) \left| \frac{d\mathbf{p}(t)}{dt} \right| \, dt : \; \mathbf{p}(t) \in C^1\left([-1,1]; \mathbb{R}^2\right), \right.$$

$$\left. \mathbf{p}(-1) = \mathbf{p}_0, \mathbf{p}(1) = \mathbf{p}_1 \right\}. \tag{5.27}$$

Remark 5.7. It is obvious to see that this metric is degenerate (i.e., zero-valued) as $\tilde{f}(\mathbf{p}) = 0$ for $\mathbf{p} = \mathbf{p}^* = \left(\rho^*, \sqrt{1 + 2c\rho^*}\right)$ and $\mathbf{p} = \mathbf{p}^{**} = \left(\rho^*, -\sqrt{1 + 2c\rho^*}\right)$. Despite such degeneracy, the infimum in (5.27) can be attained for arbitrary $\mathbf{p}_0$ and $\mathbf{p}_1$ (see Braides, 2006, Lemma 9; Wang *et al.*, 2019).

In fact, the metric $\omega(\mathbf{p}_1, \mathbf{p}_2)$ accounts for the transition costs between the profiles $\mathbf{p}_1$ and $\mathbf{p}_2$. Thus, one can deduce the energetically preferable minimisers, say $\mathbf{p}^k = (Q_{11}^k, M_1^k)$ by minimising the total transition costs that is a sum of $\omega(\mathbf{p}^k, \mathbf{p}_b(1))$, $\omega(\mathbf{p}^k, \mathbf{p}_b(-1))$. Here, $\mathbf{p}_b(1) = (-1, -1)$ and $\mathbf{p}_b(-1) = (1, 1)$ denote the boundary profiles of (Q_{11}, M_1).

According to Wang *et al.* (2019, Section 5.1), the distance $\omega(\cdot, \cdot)$ can be calculated alternatively by

$$\omega(\mathbf{p}^*, \mathbf{p}^{**}), \omega(\mathbf{p}^*, \mathbf{p}_b(1)), \omega(\mathbf{p}^{**}, \mathbf{p}_b(-1)), \omega(\mathbf{p}^*, \mathbf{p}_b(-1)), \omega(\mathbf{p}^{**}, \mathbf{p}_b(1)). \tag{5.28}$$

$$\omega\left(\mathbf{p}_0, \mathbf{p}_1\right) = \inf \left\{ \left(\int_{-1}^{1} \tilde{f}\left(\mathbf{p}(t)\right) \left| \frac{d\mathbf{p}(t)}{dt} \right|^2 \, dt \right)^{1/2} : \mathbf{p}(t) \in C^1\left([-1,1]; \mathbb{R}^2\right), \right.$$

$$\left. \mathbf{p}(-1) = \mathbf{p}_0, \mathbf{p}(1) = \mathbf{p}_1 \right\}.$$

The profiles of $\mathbf{p} = (Q_{11}, M_1)$ for each case in (5.28) are shown in Figure 5.1 which indicates that

$$\omega(\mathbf{p}^*, \mathbf{p}_b(-1)) < \omega(\mathbf{p}^{**}, \mathbf{p}_b(-1)) < \omega(\mathbf{p}^{**}, \mathbf{p}_b(1)) < \omega(\mathbf{p}^*, \mathbf{p}^{**})$$

$$< \omega(\mathbf{p}^*, \mathbf{p}_b(1)).$$

Using the computed values of those transition costs, it is clear that the OR energy minimiser converges to $\mathbf{p}^*$ almost everywhere, except close to

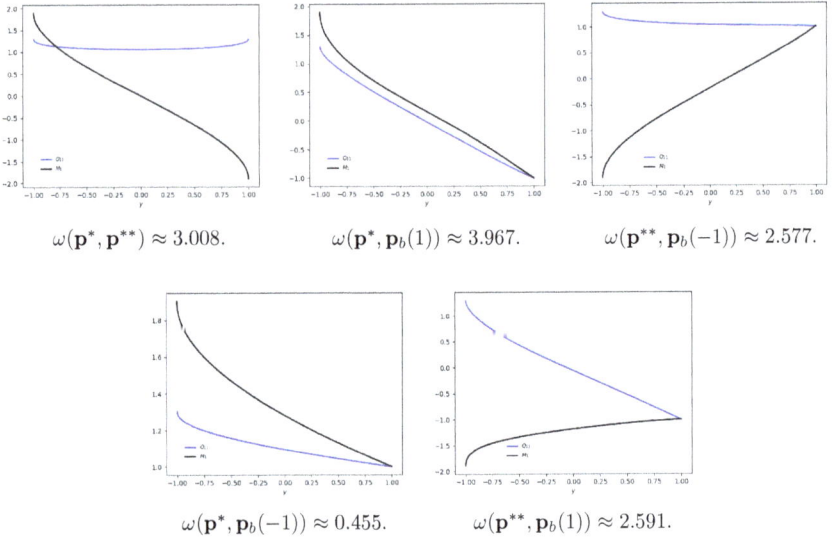

$$\omega(\mathbf{p}^*, \mathbf{p}^{**}) \approx 3.008. \qquad \omega(\mathbf{p}^*, \mathbf{p}_b(1)) \approx 3.967. \qquad \omega(\mathbf{p}^{**}, \mathbf{p}_b(-1)) \approx 2.577.$$

$$\omega(\mathbf{p}^*, \mathbf{p}_b(-1)) \approx 0.455. \qquad \omega(\mathbf{p}^{**}, \mathbf{p}_b(1)) \approx 2.591.$$

Fig. 5.1. The profiles of $\mathbf{p}$ and their corresponding transition costs in (5.28).

the boundary end points $y = \pm 1$. Moreover, no interior jumps are expected in the OR minimiser.

5.4 Summary

In this chapter, we used the LdG $\mathbf{Q}$-tensor theory to investigate the solution structure of a ferronematic problem. We introduced both the full and reduced models of ferronematics and quoted some theoretical results proven in Dalby *et al.* (2022). Our aim in the next chapters is twofold: first, verify these theoretical results computationally, and second, provide more information on the solution landscapes via numerical experiments. At first, it behoves us to introduce how to compute multiple solutions that are ubiquitous especially for solving nonlinear problems.

Chapter 6

Computing Multiple Solutions Using Deflation

Nonlinearities can make computations complicated. The occurrence of multiple solutions is one possible instance when solving a nonlinear partial differential equation. Therefore, certain efforts have been made to systematically search for multiple solutions. For instance, Shi *et al.* (2022), Yao *et al.* (2022), and Yin *et al.* (2020) utilise the saddle point dynamic theory to exploit the full solution landscape of constrained nematics liquid crystals in two dimensions; (Emerson *et al.*, 2018) adopts the deflation technique to plot disconnected bifurcations of cholesteric liquid crystals. Both approaches possess their own advantages. In this book, we shall focus on the latter choice that is based on implementations using an open-source library: Firedrake, facilitating the expressions of variational forms in coding. Therefore, we now give some simple Firedrake examples for computing LC problems based upon the Landau–de Gennes theory. Next, we summarise the deflation technique with a detailed hyperelastic example for illustrating its advantages in computing multiple solutions.

6.1 Implementation Examples

A full introduction and detailed illustration of Firedrake installation can be found in https://firedrakeproject.org/. There are further step-by-step examples given in this official website. We now give some LC-related examples in this subsection with the use of Firedrake.

6.1.1   *Two-dimensional examples*

Consider a two-dimensional domain $\Omega \subset \mathbb{R}^2$, then the symmetric, traceless tensor $\mathbf{Q}$ can be represented by

$$\mathbf{Q} = \begin{bmatrix} q_1 & q_2 \\ q_2 & -q_1 \end{bmatrix}. \tag{6.1}$$

Obviously, $\mathrm{tr}(\mathbf{Q}^3) = 0$ is always true. We use the following dimensionless energy

$$J(\mathbf{Q}) = \int_\Omega \left(\frac{1}{2}|\nabla\mathbf{Q}|^2 - \frac{c}{2}\mathrm{tr}(\mathbf{Q}^2) + \frac{cC}{4}(\mathrm{tr}(\mathbf{Q}^2))^2 \right) \mathrm{d}x, \tag{6.2}$$

where we take $\varepsilon = 1$ and $C = 10.06$ throughout the numerics in this subsection. By denoting $\mathbf{q} = (q_1, q_2)$, we derive

$$|\nabla\mathbf{Q}|^2 = 2|\nabla\mathbf{q}|^2 \quad \text{and} \quad \mathrm{tr}(\mathbf{Q}^2) = |\mathbf{Q}|^2 = 2|\mathbf{q}|^2,$$

then (6.2) becomes

$$J(\mathbf{q}) = \int_\Omega \left(|\nabla\mathbf{q}|^2 - \varepsilon|\mathbf{q}|^2 + \varepsilon C|\mathbf{q}|^4 \right) \mathrm{d}x. \tag{6.3}$$

Let $\mathbf{q} \in \mathbf{H}_g^1(\Omega) := \{\mathbf{p} \in \mathbf{H}^1(\Omega)|\mathbf{p} = \mathbf{g} \text{ on } \partial\Omega\}$ with the boundary data $\mathbf{g} \in \mathbf{H}^{1/2}(\Omega)$. The first-order optimality condition of minimising (6.3) over the admissible space $\mathbf{H}_g^1(\Omega)$ is calculated:

$$\int_\Omega 2\nabla\mathbf{q} : \nabla\mathbf{v} + \varepsilon(-2 + 4C|\mathbf{q}|^2)\mathbf{q} \cdot \mathbf{v}\mathrm{d}x = 0 \quad \forall\mathbf{v} \in \mathbf{H}_0^1(\Omega), \tag{6.4}$$

with $\mathbf{M} : \mathbf{N}$ denoting the *Frobenius* inner product for two matrices $\mathbf{M}$ and $\mathbf{N}$.

Remark 6.1. According to (6.1), we can derive that eigenvalues of $\mathbf{Q}$ satisfy

$$\lambda^2 = q_1^2 + q_2^2.$$

Before we present the numerical results in two dimensions, we specify the form of the boundary data $\mathbf{g}$. For radial alignment boundary conditions, i.e., $\mathbf{n}_0 = \mathbf{x}/|\mathbf{x}| = \left[x/\sqrt{x^2 + y^2}, y/\sqrt{x^2 + y^2} \right]^\top$, we derive the boundary condition for $\mathbf{Q}$ using the relation $\mathbf{Q}_0 = s_0(\mathbf{n}_0 \otimes \mathbf{n}_0 - \mathbf{I}_2/2)$:

$$
\mathbf{Q}_0 = \begin{bmatrix} \dfrac{x^2}{x^2+y^2} - \dfrac{1}{2} & \dfrac{xy}{x^2+y^2} \\ \dfrac{xy}{x^2+y^2} & \dfrac{y^2}{x^2+y^2} - \dfrac{1}{2} \end{bmatrix}.
$$

Here, $\mathbf{I}_2$ denotes the 2×2 identity matrix. Then, the radial boundary data of $\mathbf{q}$ is given by

$$
\mathbf{g} = \begin{bmatrix} \dfrac{x^2}{x^2+y^2} - \dfrac{1}{2}, \dfrac{xy}{x^2+y^2} \end{bmatrix}^\top. \tag{6.5}
$$

In order to visualise computed solutions, we calculate the director $\mathbf{n}$ and scalar order parameter s. First, we express $\mathbf{Q}$ in terms of $\mathbf{q} = (q_1, q_2)$ using (6.1). Since $\mathbf{Q} = s(\mathbf{n} \otimes \mathbf{n} - \mathbf{I}_2/2)$ with $\mathbf{n}$ being the eigenvector associated to the largest eigenvalue λ of tensor $\mathbf{Q}$ and $s = 2\lambda$, we can accordingly compute $\mathbf{n}$ and s by solving the corresponding eigenvalue problem of $\mathbf{Q}$.

6.1.1.1 *Numerical experiments*

An example of a two-dimensional unit circle with the radial boundary condition (6.5) is presented in Figure 6.1. It is observed that the scalar order

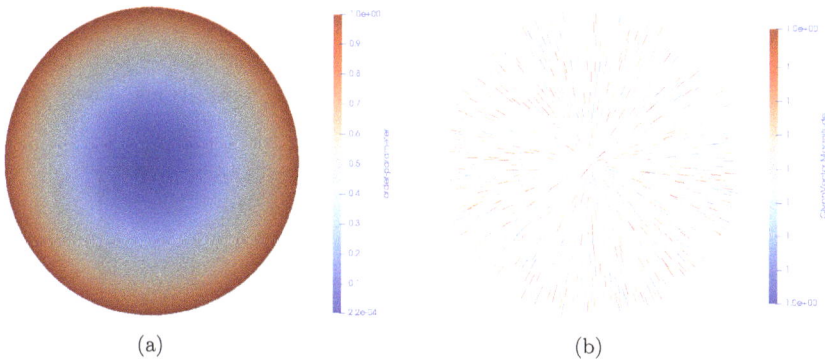

(a) (b)

Fig. 6.1. Scalar order parameter s (a) under radial boundary conditions (6.5) and its corresponding director field (b). Note that $s \approx 0$ implies the presence of defects.

parameter is approximately zero at the origin, indicating the presence of a point defect with charge $+1$.

On the other hand, if we impose tangential boundary conditions

$$\mathbf{n}_0 = \begin{cases} [0, 1]^\top & \text{on } x = 0, \\ [0, -1]^\top & \text{on } x = 1, \\ [-1, 0]^\top & \text{on } y = 0, \\ [1, 0]^\top & \text{on } y = 1, \end{cases}$$

equivalent to

$$\mathbf{g} = \begin{cases} [-1/2, 0]^\top & \text{on } x = 0 \text{ and } x = 1, \\ [1/2, 0]^\top & \text{on } y = 0 \text{ and } y = 1, \end{cases}$$

then the computed scalar order parameter on a unit square and a rectangle (with aspect ratio 1.5) is presented in Figure 6.2. It can be seen that a X-shape defect line is shown in the unit square while in the rectangle domain, it deforms into a different shape. This break of symmetry is due to the effect of the shape of domain.

6.1.2   *Three-dimensional examples*

For $\Omega \subset \mathbb{R}^3$, we can express the symmetric, traceless tensor $\mathbf{Q}$ in a form of

$$\mathbf{Q} = \begin{bmatrix} q_1 & q_3 & q_4 \\ q_3 & q_2 & q_5 \\ q_4 & q_5 & -(q_1 + q_2) \end{bmatrix}. \tag{6.6}$$

If we consider to impose Dirichlet boundary conditions, i.e., $\mathbf{Q}_0 = \mathbf{n}_0 \otimes \mathbf{n}_0 - \mathbf{I}_3/3$, then for radial boundary conditions, we have

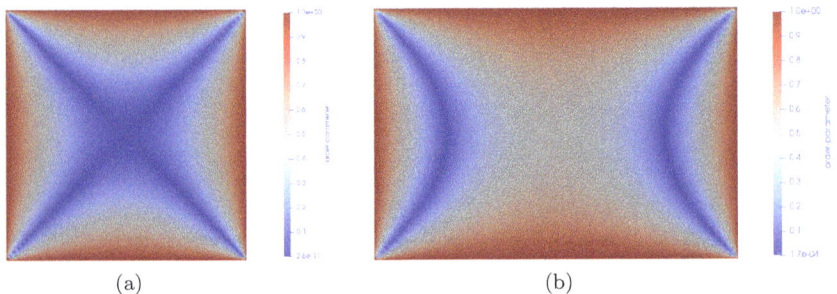

(a)   (b)

Fig. 6.2. Plot of the scalar order parameter s on a unit square (a) and a rectangle (b).

$\mathbf{n}_0 = [x/|\mathbf{x}|, y/|\mathbf{x}|, z/|\mathbf{x}|]^\top$ with $|\mathbf{x}| = \sqrt{x^2 + y^2 + z^2}$. Moreover, since

$$\mathbf{Q}_0 = \begin{bmatrix} x^2/|\mathbf{x}|^2 - 1/3 & (xy)/|\mathbf{x}|^2 & (xz)/|\mathbf{x}|^2 \\ (xy)/|\mathbf{x}|^2 & y^2/|\mathbf{x}|^2 - 1/3 & (yz)/|\mathbf{x}|^2 \\ (xz)/|\mathbf{x}|^2 & (yz)/|\mathbf{x}|^2 & z^2/|\mathbf{x}|^2 - 1/3 \end{bmatrix},$$

we have the following boundary data for $\mathbf{q}$:

$$\mathbf{g} = [x^2/|\mathbf{x}|^2 - 1/3, y^2/|\mathbf{x}|^2 - 1/3, (xy)/|\mathbf{x}|^2, (xz)/|\mathbf{x}|^2, (yz)/|\mathbf{x}|^2]^\top.$$

On the other hand, if we impose constant boundary conditions, e.g., $\mathbf{n}_0 = [0, 0, 1]^\top$, it follows that

$$\mathbf{g} = [-1/3, -1/3, 0, 0, 0]^\top.$$

Similarly, the constant boundary data $\mathbf{n}_0 = [0, 1, 0]^\top$ gives that $\mathbf{g} = [-1/3, 2/3, 0, 0, 0]^\top$ and $\mathbf{n}_0 = [1, 0, 0]^\top$ implies $\mathbf{g} = [2/3, -1/3, 0, 0, 0]^\top$.

6.1.2.1   *Visualisation*

For the purpose of visualising the solution that we obtained in numerical experiments, we calculate the eigenvalues λ_1, λ_2 and λ_3 ($\lambda_1 > \lambda_2 > \lambda_3$) of tensor $\mathbf{Q}$. Then, according to the expression (5.1):

$$\mathbf{Q} = s(\mathbf{n} \otimes \mathbf{n} - \mathbf{I}_3/3) + r(\mathbf{m} \otimes \mathbf{m} - \mathbf{I}_3/3),$$

with $\mathbf{n}, \mathbf{m}$ denoting eigenvectors corresponding to λ_1, λ_2, respectively, and $s = 2\lambda_1 + \lambda_2$, $r = \lambda_1 + 2\lambda_2$, we are able to compute $\mathbf{n}$ and s, r. For biaxial phases, these two eigenvectors $\mathbf{n}$ and $\mathbf{m}$ give two preferred directions of alignments, while order parameters r, s give the quality of alignments respectively.

In biaxial phases, we introduce another parameter β, also known as the *biaxiality parameter*, to describe the degree of biaxiality and it is defined as

$$\beta(Q) = 1 - 6\frac{(\mathrm{tr}(\mathbf{Q}^3))^2}{(\mathrm{tr}(\mathbf{Q}^2))^3}. \tag{6.7}$$

By Majumdar and Zarnescu (2010, Lemma 1), $\beta(\mathbf{Q}) \in [0, 1]$ and $\beta(\mathbf{Q}) = 0$ if and only if $\mathbf{Q}$ is uniaxial. Moreover, when $\beta(\mathbf{Q}) = 1$, this indicates the maximal biaxiality (i.e., $r/s = 1/2$) is achieved.

6.1.2.2 *Numerical experiments*

In this subsection, we present some three-dimensional numerical results. An example of imposing radial boundary conditions in a unit ball is given in Figure 6.3. It is shown that all points except those at the origin are uniaxial, while the origin of the ball appears to have a +1 point defect with the biaxiality parameter being around 0.35. In such computation, we observe that the point defect is actually biaxial rather uniaxial.

Another example of three-dimensional numerical tests is followed in Figure 6.4, where the domain is a unit cube. The derived defect at $z = 0.5$ is also denoted as *plane defect* in Nochetto *et al.* (2017, Section 5.3). Our result is consistent with the numerical results in Nochetto *et al.* (2017): the defect plane appears at $z = 0.5$. Moreover, we find that the biaxiality parameter of the defect plane $z = 0.5$ is around zero according to Figure 6.4.

In the next example, we show the result of *propeller defect* in a cube (Nochetto *et al.*, 2017, Section 5.4) in Figure 6.5. Here, we impose Dirichlet boundary conditions

$$\mathbf{n}_0 = \frac{[x - 0.5, y - 0.5, 0]^\top}{\sqrt{(x - 0.5)^2 + (y - 0.5)^2}}$$

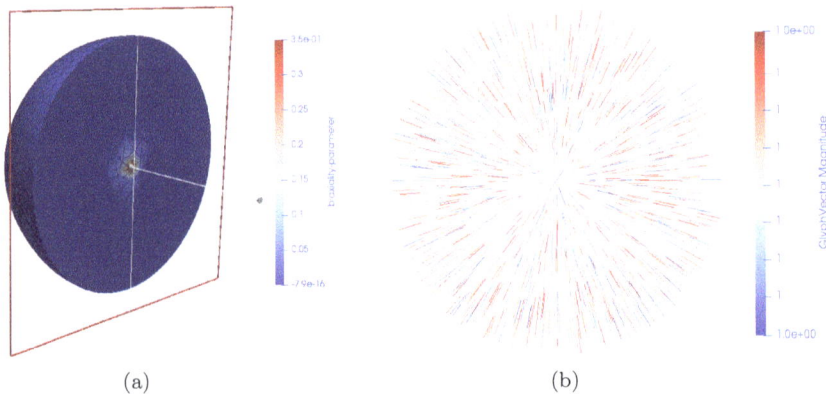

(a)   (b)

Fig. 6.3. A clip (along $y = 0$) of the biaxiality parameter β (a) and director (b). Notice that the point defect at the center has the value of $\beta \approx 0.35$, implying that this +1 defect is actually biaxial rather uniaxial.

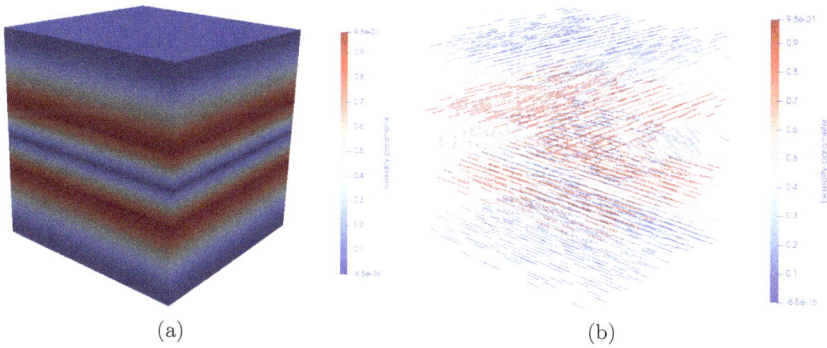

(a)   (b)

Fig. 6.4. Three-dimensional plot of the biaxiality parameter β (a) and director (b) in a unit cube. There is a plane defect at $z = 0.5$ having $\beta \approx 0$ and it is surrounded by the plane of $\beta \approx 0.98$.

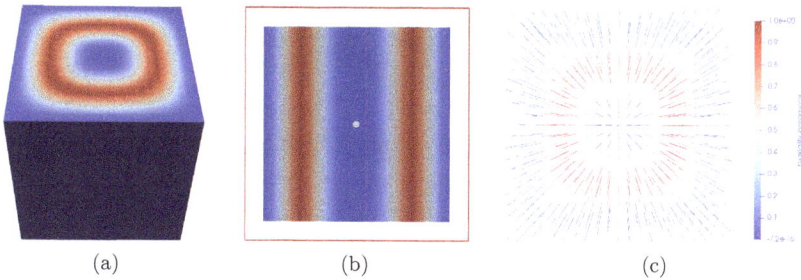

(a)   (b)   (c)

Fig. 6.5. Results of nematic liquid crystals in a unit cube with planar radial boundary conditions (6.8): the biaxiality parameter β (a), a clip (along $x = 1/2$) of β (b) and the director on the plane of $z = 1$ (c). Notice that a point defect ($\beta \approx 0$) is presented in the rightmost picture; again, it is surrounded by points that have $\beta \approx 1$.

for all four vertical sides of the cube, i.e., $\{x = 0\} \cup \{x = 1\} \cup \{y = 0\} \cup \{y = 1\}$. Therefore, the boundary data for $\mathbf{q}$ is

$$\mathbf{g} = \left[\frac{(x - 0.5)^2}{(x - 0.5)^2 + (y - 0.5)^2} - \frac{1}{3}, \frac{(y - 0.5)^2}{(x - 0.5)^2 + (y - 0.5)^2} - \frac{1}{3}, \right.$$
$$\left. \frac{(x - 0.5)(y - 0.5)}{(x - 0.5)^2 + (y - 0.5)^2}, 0, 0 \right]^{\top}. \tag{6.8}$$

One should notice that (Nochetto *et al.*, 2017, Section 5.4) applies the one-constant Ericksen model which only takes uniaxial liquid crystals into account. In our case, we allow the presence of biaxiality and Figure 6.5

shows us that indeed the biaxiality appears around the central vertical axis.

6.2 Deflation[1]

Above examples show the potential and convenience of implementing LC problems in Firedrake. During such experience, it is expected to observe multiple solutions for a nonlinear problem. Therefore, we introduce the idea of deflation and provide implementation examples below. Consider a parameter-dependent nonlinear problem

$$f(u, \lambda) = 0 \quad \text{for } u \in U \text{ and } \lambda \in \mathbb{R}, \tag{6.9}$$

where U is an admissible space for u and λ is the parameter. For our purposes, we assume that this problem permits multiple solutions for some values of λ, which we wish to find. Its bifurcation diagram will then visualise how solutions change as the parameter λ varies over the range $[\lambda_{\min}, \lambda_{\max}]$.

A classical strategy used in solving (6.9) is a combination of *arc-length continuation* and *branch switching*. Briefly speaking, given an initial guess (u_0, λ_0) on a branch, arc-length continuation will trace out the remaining part of this branch along the variations of the parameter λ. On the other hand, branch switching will detect bifurcation points along the branch and construct initial solutions on branches emanating from it. Once one solution on each emanating branch is computed, arc-length continuation is applied to complete the remaining part of the new branch.

This combination of arc-length continuation and branch switching is very powerful for computing connected bifurcation diagrams. However, in the presence of geometric imperfections that disconnect branches, it fails to compute these other branches. One approach is to restore the broken symmetry group, find all branches of the now continuous diagram via branch switching, re-introduce the asymmetry, and continue the solutions from the symmetric to the asymmetric state. If the asymmetry is introduced via a parameter in the equations (e.g., asymmetry in the loading), this is straightforward, but if the asymmetry is introduced in other ways (such as in the geometry) this procedure can be very difficult to apply. Farrell

---

[1]This section and the hyperelastic example following on are adapted from the author's work in Xia *et al.* (2020).

et al. (2016) remedy this issue by introducing the deflated continuation algorithm to discover disconnected branches from known ones without requiring any detection of the bifurcation point. The algorithm is used in this work for computing the bifurcation diagrams of different stiffener types.

At the heart of the deflated continuation method is the deflation technique (Farrell *et al.*, 2015). Historically, the deflation technique was first applied to finding distinct solutions to scalar polynomials (Wilkinson, 1994). Brown and Gearhart (1971) then extended this deflation approach to solving systems of nonlinear algebraic equations via the construction of deflation matrices. A more recent study of Farrell *et al.* (2016) extended the deflation technique to the case of infinite-dimensional Banach spaces, appropriate for partial differential equations. In the following, we recall the idea of deflation.

For a fixed parameter λ^*, the parameter-dependent problem (6.9) becomes

$$F(u) := f(u, \lambda^*) = 0. \tag{6.10}$$

Suppose that (6.10) permits multiple solutions and the Newton iteration converges to a known solution u^*. The goal of deflation is to find as many solutions to $F(u) = 0$ in a way that the Newton iteration will never converge to known solutions even with the same initial guess. To this end, a new problem

$$G(u) := M(u; u^*)F(u) = 0$$

is constructed, where $M(u; u^*)$ is a *deflation operator* and G satisfies the following two properties:

(1) $G(u) = 0$ has the same solutions as those of $F(u) = 0$; that is to say, for all $u \neq u^*$, $F(u) = 0 \iff G(u) = 0$.
(2) For a known solution u^* to $F(u) = 0$, G will not converge to u^* again; i.e., given any sequence $u_i \to u^*$, $\liminf_{u_i \to u^*} \|G(u_i)\| > 0$.

The form of $M(u; u^*)$ used in this work is the shifted deflation operator

$$M(u; u^*) = \frac{1}{\|u - u^*\|^p} + \alpha, \tag{6.11}$$

where the pole strength p governs the rate at which the function approaches the introduced singularity, and the shift parameter α ensures that the deflated problem recovers the behaviour of the original problem far from previously found solutions as $\|u - u^*\| \to \infty$. In our algorithm, the values $p = 2$ and $\alpha = 1$ are adopted.

We now give a brief description of the deflated continuation algorithm. The algorithm proceeds by continuation over a range of values of ε. Consider the step in the algorithm going from $\varepsilon = \varepsilon^-$ to $\varepsilon = \varepsilon^+$. Suppose that n solutions $u_1^-, u_2^-, \ldots, u_n^-$ are known at $\varepsilon = \varepsilon^-$. The step proceeds in two phases. First, each solution u_i^- is continued from ε^- to ε^+ to yield u_i^+ (using arclength, tangent or standard continuation).[2] As each solution u_i^+ is computed, it is deflated away from the nonlinear problem at $\varepsilon = \varepsilon^+$. Once all known solutions have been continued, the search phase of the algorithm begins. Each previous solution u_i^- is used again as initial guess for the nonlinear problem at $\varepsilon = \varepsilon^+$; the deflation operator ensures that the solve will not converge to any of the known solutions u_i^+, and hence if Newton's method converges it must converge to a new, unknown solution. Importantly, this unknown solution may lie on a disconnected branch. If an initial guess yields a new branch, the new solution is deflated and the initial guess used repeatedly until failure. Once all initial guesses from ε^- have been exhausted, the step completes and the algorithm proceeds to the next step. This is repeated until the continuation parameter reaches a desired target value. The search is applied at all steps, i.e., no *a priori* knowledge of the location of the disconnected bifurcations is assumed. For more details, including application to standard benchmark cases, see Farrell *et al.* (2016).

6.3   A Hyperelastic Example

6.3.1   *Saint Venant–Kirchhoff hyperelastic model*

Consider a three-dimensional body occupying a *reference configuration* $\mathcal{B}_0$ with Lipschitz continuous boundary Γ subject to certain loads to the body, thus leading to a *deformed configuration* $\mathcal{B}$. In this model, we characterise the deformation by the displacement $\mathbf{u} : \mathcal{B}_0 \to \mathbb{R}^3$ and define the *deformation gradient* tensor by

$$\mathbf{F}(\mathbf{x}) = \mathbf{I} + \nabla \mathbf{u}(\mathbf{x}),$$

---

[2]In the problems considered here standard zero-order continuation is sufficient, so we use this.

where $\mathbf{I}$ is the identity second-order tensor. For Saint Venant–Kirchhoff hyperelastic materials, the constitutive equation (i.e., the stress-strain relation) can be written as

$$\boldsymbol{\sigma}(\mathbf{E}) = \lambda(\mathrm{tr}\mathbf{E})\mathbf{I} + 2\mu\mathbf{E}, \qquad (6.12)$$

with λ and μ being *Lamé parameters*, $\boldsymbol{\sigma}$ the *second Piola–Kirchhoff stress* tensor and $\mathbf{E}$ the *Lagrangian strain* tensor given by

$$\mathbf{E} = (\mathbf{F}^T\mathbf{F} - \mathbf{I})/2.$$

Remark 6.2. In the implementation, Lamé parameters are determined by *Young's modulus E* and *Poisson's ratio* ν:

$$\lambda = \frac{E\nu}{(1 - 2\nu)(1 + \nu)} \quad \text{and} \quad \mu = \frac{E}{2(1 + \nu)}.$$

In order to have a well-posed problem, additional boundary conditions are needed. We divide the boundary Γ into two disjoint parts:

$$\Gamma = \Gamma_D \cup \Gamma_N,$$

with the Dirichlet boundary Γ_D and the traction boundary Γ_N. On the top and bottom boundary faces $\Gamma_D = \Gamma_{\text{top}} \cup \Gamma_{\text{bottom}}$, we enforce

$$\mathbf{u} = \mathbf{u}_0 \quad \text{on } \Gamma_D, \qquad (6.13)$$

where

$$\mathbf{u}_0 = \begin{cases} [0,0,0]^T & \text{on } \Gamma_{\text{top}}, \\ [0,0,\varepsilon]^T & \text{on } \Gamma_{\text{bottom}}, \end{cases} \qquad (6.14)$$

with ε being a parameter that will be continued in the deflated continuation algorithm (Farrell *et al.*, 2016). Note that the z-direction corresponds to the longitudinal (axial) direction of stiffeners in this work. For simplicity, we assume that the aircraft stiffeners are homogeneous, isotropic and frame-indifferent (Ciarlet, 1988).

Remark 6.3. The displacement-controlled boundary condition on Γ_D guarantees that all degrees of freedom at Γ_D have the same displacement condition. Imposing a traction on the bottom boundary may not yield an even displacement distribution.

Remark 6.4. Note that ε in the boundary data $\mathbf{u}_0$ corresponds to the axial displacement applied to the stiffeners.

On Γ_N, we have no traction, i.e.,

$$\mathbf{t}(\mathbf{u}) = \mathbf{0} \quad \text{on } \Gamma_N, \tag{6.15}$$

where the traction $\mathbf{t}(\mathbf{u})$ is defined by

$$\mathbf{t}(\mathbf{u}) = \mathbf{P}(\mathbf{u})\,\mathbf{n},$$

with $\mathbf{P} = \boldsymbol{\sigma}\mathbf{F}$ denoting the *first Piola–Kirchhoff stress* tensor and $\mathbf{n}$ the outward normal to the boundary surface. In addition, the uniaxial applied force $\mathbf{f}_{\text{ext}}$ can be calculated from the second Piola–Kirchhoff stress tensor $\boldsymbol{\sigma}$ via

$$\mathbf{f}_{\text{ext}} = -\int_{\Gamma_{\text{bottom}}} \mathbf{n} \cdot [\boldsymbol{\sigma}(\mathbf{u})\,\mathbf{n}]\ ds. \tag{6.16}$$

Here, the negative sign is added to represent the positive compressive force. Then, it is known that the average stress over the bottom face is computed by

$$\boldsymbol{\sigma} = \frac{\mathbf{f}_{\text{ext}}}{|\Gamma_{\text{bottom}}|}, \tag{6.17}$$

with $|\Gamma_{\text{bottom}}|$ denoting the measure of the bottom face.

The boundary value problem considered in this work is

$$-\nabla \cdot \mathbf{P}(\mathbf{u}) = \mathbf{b} \quad \text{in } \mathcal{B}_0,$$

$$\mathbf{u} = \mathbf{u}_0 \quad \text{on } \Gamma_D, \tag{6.18}$$

$$\mathbf{t}(\mathbf{u}) = \mathbf{0} \quad \text{on } \Gamma_N,$$

with $\mathbf{b}$ being the body force vector. In the implementation, we let $\mathcal{B}_0$ be the aircraft stiffener and ignore the gravitational body force, i.e, $\mathbf{b} = \mathbf{0}$, as it is negligible compared with the compressive force that we impose.

Denote the admissible function space of the displacement by

$$V = W^{1,4}(\mathcal{B}_0; \mathbb{R}^3).$$

The weak form of (6.18) can be derived as: find $\mathbf{u} \in V$ satisfying $\mathbf{u} = \mathbf{u}_0$ on Γ_D such that

$$R(\mathbf{u}, \mathbf{v}) := \int_{\mathcal{B}_0} \mathbf{P}(\mathbf{u}) : \nabla\mathbf{v} = 0, \tag{6.19}$$

for all $\mathbf{v} \in V$ satisfying $\mathbf{v} = \mathbf{0}$ on Γ_D. The Dirichlet boundary condition $\mathbf{u} = \mathbf{u}_0$ will be enforced weakly later using Nitsche's method.

Remark 6.5. The $W^{1,4}$-regularity is needed to make the weak form (6.19) well-defined. Indeed, by direct computations, we can obtain

$$\mathbf{P}(\mathbf{u}) = \lambda \left(\nabla \cdot \mathbf{u} + \frac{|\nabla \mathbf{u}|^2}{2} \right) (\mathbf{I} + \nabla \mathbf{u}) + \mu (\nabla \mathbf{u} + \nabla \mathbf{u}^T + \nabla \mathbf{u}^T \nabla \mathbf{u})(\mathbf{I} + \nabla \mathbf{u}).$$

If $\mathbf{u}, \mathbf{v} \in W^{1,p}$, then $\mathbf{P}(\mathbf{u}) \in L^{p/3}$ and $\nabla \mathbf{v} \in L^p$. Thus, $\mathbf{P}(\mathbf{u}) : \nabla \mathbf{v}$ is in $L^{p/4}$. This requires $p = 4$ at least for (6.19) to be well-defined. Moreover, by the Sobolev embedding theorem (Ciarlet, 1988, Theorem 6.1-3), the $W^{1,4}$-regularity guarantees that pointwise evaluation is well-defined.

6.3.2 *Enforcement of the essential boundary condition*

The traction-free boundary condition (6.15) is naturally enforced in (6.19) via the divergence theorem; it remains to enforce the essential boundary condition (6.14). Throughout this work, we will follow Nitsche's method (Nitsche, 1971) to weakly impose the Dirichlet boundary condition $\mathbf{u} = \mathbf{u}_0$ on Γ_D. To this end, we add the following two terms

$$\gamma \int_{\Gamma_D} (\mathbf{u} - \mathbf{u}_0) \cdot \mathbf{v} \, \mathrm{dx} - \int_{\Gamma_D} \mathbf{t}(\mathbf{u}) \cdot \mathbf{v} \, \mathrm{ds} \tag{6.20}$$

to the weak form (6.19). Here, $\gamma > 0$ is a large penalty parameter, necessary for numerical stability (Rüberg *et al.*, 2016). Note that the second term in (6.20) arises from integration by parts of the divergence of the first Piola–Kirchhoff stress tensor $\mathbf{P}$ in (6.18).

Remark 6.6. The term $\int_{\Gamma_D} \mathbf{t}(\mathbf{u}) \cdot \mathbf{v} \, \mathrm{ds}$ in (6.20) is well-defined. Indeed, this can be seen from the inequality (Embar *et al.*, 2010; Lu *et al.*, 2019)

$$\|\mathbf{P}(\mathbf{u})\, \mathbf{n}\|_{0,\Gamma_D}^2 \leq c \int_{\mathcal{B}_0} \mathbf{P}(\mathbf{u}) : \nabla \mathbf{u} \, \mathrm{dx} \quad \forall \mathbf{u} \in V,$$

with $c > 0$ a mesh-dependent constant.

Consequently, we summarise the final variational problem used in this work as follows: find $\mathbf{u} \in V$ such that

$$R^*(\mathbf{u}, \mathbf{v}) := R(\mathbf{u}, \mathbf{v}) + \gamma \int_{\Gamma_D} (\mathbf{u} - \mathbf{u}_0) \cdot \mathbf{v} \, \mathrm{dx} - \int_{\Gamma_D} \mathbf{t}(\mathbf{u}) \cdot \mathbf{v} \, \mathrm{ds} = 0 \quad \forall \mathbf{v} \in V. \tag{6.21}$$

Essentially, (6.21) is a consistent formulation as the additional penalty term is zero for an exact solution.

Remark 6.7. Here, we use the non-symmetric version of Nitsche's method (Freund and Stenberg, 1995) for ease of the stability analysis (see A.1).

Furthermore, we can see that the variational problem (6.21) is nonlinear due to the presence of the nonlinear stress–strain relation (6.12). Hence, the classical Newton method is applied and the rth Newton iteration takes the form of

$$DR^*(\mathbf{u}^r, \mathbf{v})[\delta\mathbf{u}] = -R^*(\mathbf{u}^r, \mathbf{v}), \qquad (6.22)$$

with the update $\delta\mathbf{u} \in V$ to the current approximation $\mathbf{u}^r$. Here, the first Gâteaux derivative is given by

$$DR^*(\mathbf{u}^r, \mathbf{v})[\delta\mathbf{u}] = Da(\mathbf{u}^r, \mathbf{v})[\delta\mathbf{u}] - \int_{\Gamma_D} Dt(\mathbf{u})[\delta\mathbf{u}] \cdot \mathbf{v} \, ds + \gamma \int_{\Gamma_D} \delta\mathbf{u} \cdot \mathbf{v} \, ds,$$

where

$$Da(\mathbf{u}^r, \mathbf{v})[\delta\mathbf{u}] := \int_\Omega \nabla\delta\mathbf{u} : \mathbf{C}(\mathbf{u}^r) : \delta\mathbf{v} \, dx,$$

with the *fourth-order elasticity* tensor $\mathbf{C}$ in the form of

$$\mathbf{C}_{ijkl}(\mathbf{u}) = \left(\nabla \cdot \mathbf{u} + \frac{|\nabla\mathbf{u}|^2}{2} \right) \delta_{ik}\delta_{jl} + (\mathbf{I} + \nabla\mathbf{u})_{ij}(\mathbf{I} + \nabla\mathbf{u})_{kl}$$

$$+ (\nabla\mathbf{u} + \nabla\mathbf{u}^T + \nabla\mathbf{u}^T\nabla\mathbf{u})_{ik}\delta_{jl} + (\mathbf{I} + \nabla\mathbf{u})_{ik}(\mathbf{I} + \nabla\mathbf{u})_{jl}$$

$$+ (\mathbf{I} + 2\nabla\mathbf{u} + \nabla\mathbf{u}\nabla\mathbf{u})_{il}\delta_{jk}.$$

Here, δ_{ij} is the *Kronecker delta* defined by

$$\delta_{ij} = \begin{cases} 0 & \text{if } i \neq j, \\ 1 & \text{if } i = j, \end{cases}$$

and A_{ij} denotes the (i, j)th entry of the second-order tensor A.

In the implementation, we take the parameters $E = 69$ GPa, $\nu = 0.334$ for aluminium stiffeners (The Engineering ToolBox, 2003, 2008) and choose $\gamma = 10^{15}$ based on unreported preliminary experiments. The choice of γ is related to the stability analysis of the Newton system (6.22), discussed in A.1.

Further discussions of the Saint Venant–Kirchhoff model and other models can be found in Bower (2009) and Ciarlet (1988).

As discussed in the previous section, more robust methods investigating multiple bifurcation paths should be considered. We notice that due to the presence of nonlinearity in the Saint Venant–Kirchhoff model (6.21) and possibly geometric imperfections in our concerned aircraft stiffeners, the variational problem (6.21) can permit multiple equilibrium states. In this section, we will briefly review the deflation technique and use the deflated continuation algorithm proposed in Farrell *et al.* (2016) to exploit the buckling profiles for three practical types of stiffeners (see Figure 6.6).

Three practical aircraft stiffeners are considered: the L-shaped asymmetric type, the L-shaped symmetric type and the Z-shaped asymmetric type, as detailed in Figure 6.6. It is important to emphasise that, in the present study, the nonlinear post-buckled bifurcation paths are investigated for the stiffener profiles in a nonassembled configuration, i.e., not as part of a stiffened panel. In real applications, the stiffeners would be assembled in a panel, where post-buckling configurations of such stiffeners can usually be achieved before the ultimate loads supported by the structure.

All stiffeners are modelled with Saint Venant–Kirchhoff hyperelastic materials and used to test the algorithm under displacement-controlled uniaxial compression.

6.3.3 *Bifurcation analysis of buckling behaviours*

In this section, the deflated continuation algorithm is applied to investigate the buckling behaviour of three different aircraft stiffeners, i.e., the L-shaped asymmetric profile, the L-shaped symmetric profile and the Z-shaped asymmetric profile (see Figure 6.6). We perform a uniaxial compression test along the z-axis, with the compressive force applied to the bottom face of each stiffener. SI units are adopted for all physical quantities in the subsequent experiments.

Throughout the simulations, the boundary data ε is continued with the continuation step $\Delta\varepsilon = 10^{-5}$. All numerical experiments are based on a continuous piecewise linear discretisation of the displacement function space. For the linearisation, we employ Newton's method with the L^2 line search algorithm of PETSc (Balay *et al.*, 2018) with relative and absolute tolerances 10^{-7}. The solve is terminated with failure if convergence

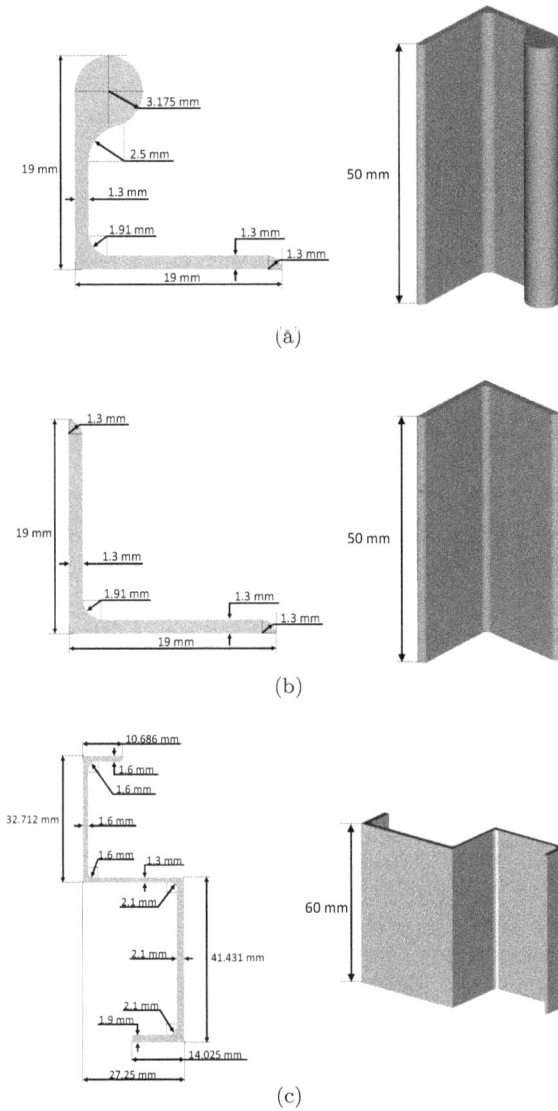

Fig. 6.6. Aircraft stiffener profiles used in this work and their corresponding geometries. (a) L-shaped asymmetric stiffener; (b) L-shaped symmetric stiffener; (c) Z-shaped asymmetric stiffener.

is not achieved in 50 nonlinear iterations. At each Newton iteration, the linearised system is solved by GMRES with a V-cycle multigrid preconditioner, where the coarse grid problem is solved by Cholesky factorisation and the additive block Successive Over-Relaxation (SOR) algorithm is used as relaxation (Balay *et al.*, 2018). The computations are performed on eight cores, parallelised using MPI.

6.3.3.1 *L-shaped asymmetric profile*

In this experiment, the boundary data ε is varied in the range $[0, 0.002]$. Figure 6.7 illustrates the bifurcation diagram of the functional $\mathbf{u}_1(0.019, 0, 0.025)$ with respect to the parameter ε. For $\varepsilon \lesssim 0.00078$, there is only one solution to the problem (6.21); two more solutions appear until $\varepsilon \approx 0.00086$, after which there exist at least five solutions. Then from $\varepsilon \approx 0.00139$, two more branches are found, leading to a total number of seven solutions discovered.

The resulting seven solutions at $\varepsilon = 0.002$ are given in Figure 6.8. For a better connection with the bifurcation diagram in Figure 6.7, we point out that the first and second deformed profiles in Figure 6.8 correspond to the lowest and uppermost branch.

Furthermore, we compute the stability through calculating the inertia of the Hessian matrix of the energy function $\Phi(\mathbf{u})$ with a Cholesky

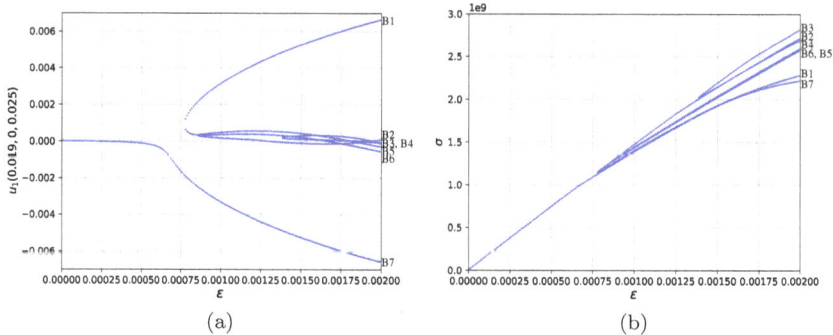

(a)   (b)

Fig. 6.7. (a) The bifurcation diagram of the L-shaped asymmetric stiffener where the functional $\mathbf{u}_1(0.019, 0, 0.025)$ corresponds to the y-component of the displacement evaluated at the midpoint of the left boundary. (b) The average stress at the bottom face. The enumeration of the branches from B1 to B7 corresponds to the images shown in Figure 6.8.

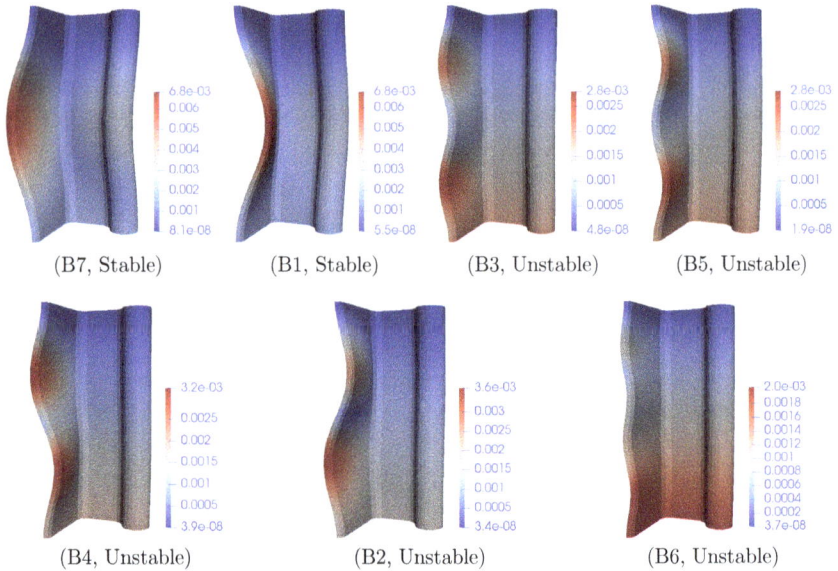

(B7, Stable)   (B1, Stable)   (B3, Unstable)   (B5, Unstable)

(B4, Unstable)   (B2, Unstable)   (B6, Unstable)

Fig. 6.8. Seven buckling modes of the L-shaped asymmetric stiffener at $\varepsilon = 0.002$. The colours refer to the magnitude of the displacement from the original configuration.

factorisation (Nocedal and Wright, 1999, Section 16.2). This reveals that the first two buckling profiles in Figure 6.8 are stable (i.e., the Hessian matrix is positive definite) while the remaining five modes (with a nonzero number of negative eigenvalues, making the Hessian matrix indefinite) are unstable. These five unstable buckling profiles can be easily perturbed.

From Figure 6.7, it is noticeable that there exist disconnected branches even though the body force and the traction are zero in the model. This is due to the nonsymmetric geometry of the aircraft stiffener, making it easier to buckle outwards than inwards.

Additionally, we plot the average stress over the bottom face computed by (6.17) in Figure 6.7. It is shown that for sufficiently small deformations, it is proportional to the displacement, as expected from Hooke's law. For relatively large deformations, their relationship becomes nonlinear. Notice that the yielding stress of Aluminum is 70 MPa (Huang and Asay, 2005) and our obtained stress is about the level of 1000 MPa, as can be seen from Figure 6.7. Other mathematical models (Simo and Hughes, 2000) that include plasticity should therefore be considered in the future.

Remark 6.8. One might wonder about the utility of identifying the unstable buckling modes presented in Figure 6.8 above and Figures 6.10 and 6.12 below, as only stable solutions can be physically observed in experiments. However, unstable solutions provide important information about the energy barrier that the system must overcome to switch from one stable solution to another. Of all possible paths in the energy landscape, the one with lowest energy cost will go through one of those unstable solutions (a mountain pass). This intuitive statement is formalised by the so-called *Mountain Pass Theorem*, see Evans (2010, Section 8.5). Therefore, knowledge of the unstable modes gives knowledge of the energetic stability of the different local minimisers.

6.3.3.2 *L-shaped symmetric stiffener*

We conduct similar numerical experiments for the L-shaped symmetric stiffener which possesses a geometric symmetry due to the absence of the bulb (see Figure 6.6). In our preliminary experiments, we observe many more branches in the bifurcation diagram for $\varepsilon \in [0, 0.002]$. To make a clear bifurcation diagram, we instead illustrate the case of varying ε in $[0, 0.0007]$.

The bifurcation diagram of the functional $\mathbf{u}_1(0.019, 0, 0.025)$ is shown in Figure 6.9. We first observe that when $\varepsilon \lesssim 0.00031$, there exists only one solution. The system then undergoes a pitchfork bifurcation with three solutions until $\varepsilon \approx 0.00047$, after which it presents five solutions. Around

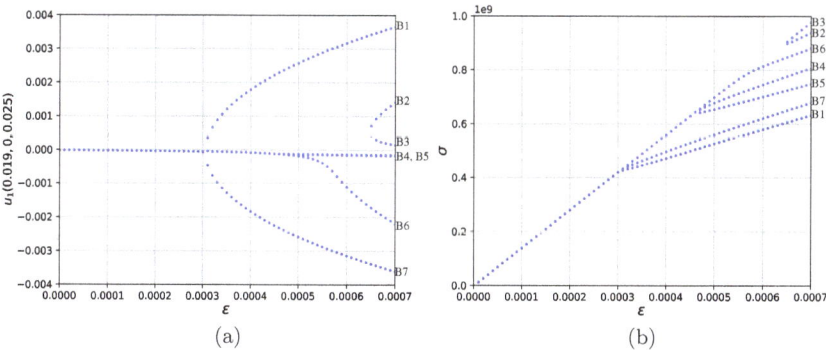

(a)   (b)

Fig. 6.9. (a) The bifurcation diagram of the L-shaped symmetric stiffener where the functional $\mathbf{u}_1(0.019, 0, 0.025)$ is the y-component of the displacement at the midpoint $(0.019, 0, 0.025)$ of the left boundary. (b) The average stress at the bottom face. The enumeration of the branches from B1 to B7 corresponds to the images shown in Figure 6.10.

the point of $\varepsilon \approx 0.00065$, it starts to buckle in seven different modes with two new disconnected branches.

One expects a connected bifurcation diagram for this stiffener profile because of its geometric symmetry. However, Figure 6.9 reveals a disconnected bifurcation around $\varepsilon \approx 0.00065$. This is the correct diagram for this discrete problem, and the reason is subtle: while the mesh is almost perfectly symmetric, there is a slight asymmetry around the centre web. In general, exactly preserving the continuous symmetries of the geometry during mesh generation is very difficult. Even small perturbations to the symmetry can lead to a (discrete) disconnected bifurcation diagram that may not be easily captured by conventional arc-length continuation and branch switching algorithms. The deflated continuation algorithm we used helps us capture the relevant branches without needing to enforce symmetry of the mesh, improving the flexibility of the computations.

In this numerical experiment, we have found seven buckling modes in total and they are all illustrated in Figure 6.10 in pairing order. There are three $\mathbb{Z}_2$-symmetric pairs of modes, as well as the single $\mathbb{Z}_2$-symmetric

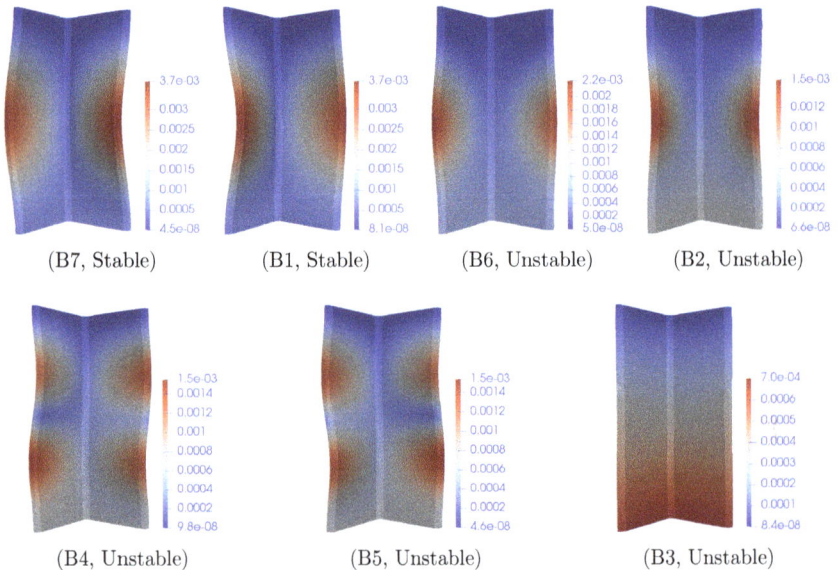

| (B7, Stable) | (B1, Stable) | (B6, Unstable) | (B2, Unstable) |

| (B4, Unstable) | (B5, Unstable) | (B3, Unstable) |

Fig. 6.10. Seven buckling modes of the L-shaped symmetric stiffener at $\varepsilon = 0.0007$. The colours refer to the magnitude of the displacement.

compressed state. Additionally, the stability of each buckling profile is indicated in Figure 6.10. We can see that only the first two profiles are stable while the remaining five buckling profiles can be easily perturbed.

The average stress over the bottom face computed by (6.17) is plotted in Figure 6.9. We can observe a linear stress-strain relation for small ε corresponding to Hooke's law and then this relation becomes nonlinear for larger deformations. As before, a more physically realistic model should incorporate plasticity.

6.3.3.3 *Z-shaped asymmetric stiffener*

For the Z-shaped asymmetric stiffener profile, its bifurcation diagram is shown in Figure 6.11. To keep the number of solutions considered manageable, we consider $\varepsilon \in [0, 0.0027]$. The disconnection of the bifurcation comes again from the asymmetry of the domain, similar to the case of the L-shaped asymmetric stiffener. It can be seen that the diagram starts to bifurcate at $\varepsilon \approx 0.00191$, obtaining three solutions, and approximately at $\varepsilon = 0.00203$, five branches appear until $\varepsilon \approx 0.00249$ where four more solutions are found. Consequently, there are nine branches in total and Figure 6.12 illustrates these buckling profiles at $\varepsilon = 0.0027$. Essentially, four pairs of buckling modes have been discovered, along with the neutrally compressed state.

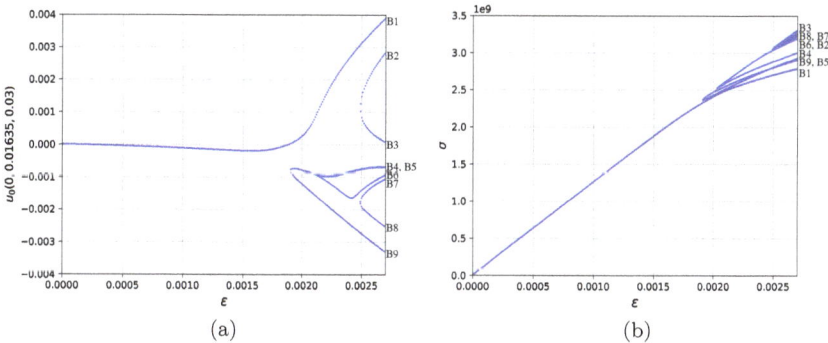

(a)   (b)

Fig. 6.11. (a) The bifurcation diagram of the Z-shaped asymmetric stiffener where the functional $\mathbf{u}_0(0, 0.01635, 0.03)$ is taken to be the x-component of the displacement at the centre (0, 0.01635, 0.03) of the flange. (b) The average stress at the bottom face. The enumeration of the branches from B1 to B9 corresponds to the images shown in Figure 6.12.

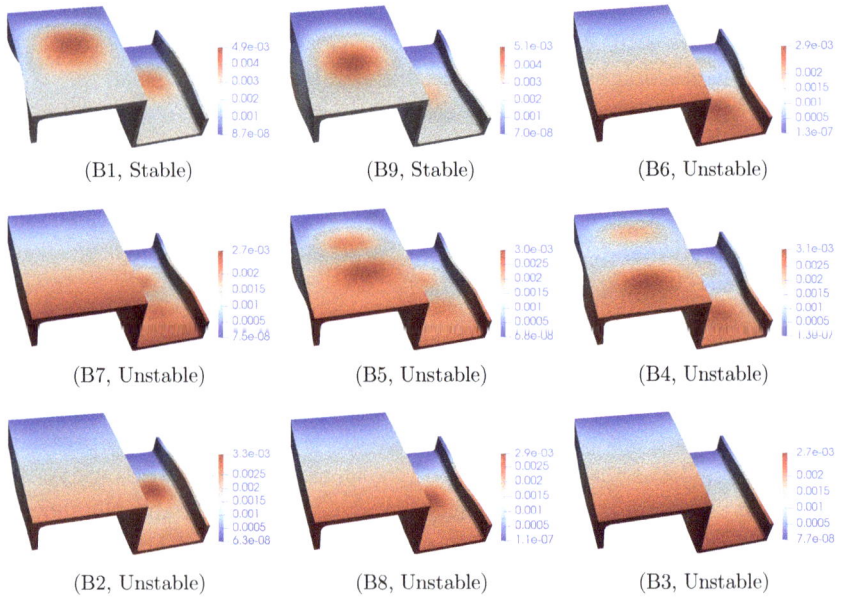

| (B1, Stable) | (B9, Stable) | (B6, Unstable) |

| (B7, Unstable) | (B5, Unstable) | (B4, Unstable) |

| (B2, Unstable) | (B8, Unstable) | (B3, Unstable) |

Fig. 6.12. Nine buckling modes for the Z-shaped asymmetric stiffener at $\varepsilon = 0.0027$. The colours refer to the magnitude of the displacement.

We also point out that the uppermost and lowest branch in Figure 6.11 correspond to the first and the second buckling profiles in Figure 6.12. This implies that the Z-shaped asymmetric stiffener is easier to buckle upwards rather than downwards.

Regarding the stability of each buckling profile, it is shown that only the first two buckling profiles in Figure 6.12 are stable (i.e., the Hessian matrix is positive definite). The remaining seven unstable solutions in Figure 6.12 can be easily perturbed.

The average stress over the bottom face, computed by (6.17), is plotted in Figure 6.11. The linear stress-strain relation for small ε is also observed, which again verifies Hooke's law, and again indicates that plasticity should be considered to achieve more physically realistic results.

6.4   Summary

In this chapter, we illustrate some examples of using Firedrake and deflation techniques to explore multiple solutions naturally arising from nonlinear problems. Some common two- and three-dimensional nematic examples are first given. Then, in order to find as many solutions as possible, we summarise the deflation technique combined with arc-length continuation algorithm. A hyperelastic example is shown in detail, elucidating the validity of the use of the deflated continuation algorithm in computing multiple solutions. In the next chapter, we will use this algorithm to solve the ferronematic problem introduced in Chapter 5.

Chapter 7

Numerical Verifications for Ferronematics

In this chapter, we perform numerical experiments to validate the theoretical results proven in Dalby *et al.* (2022) and quoted in Chapter 5, and understand the interplay between the elastic constant k and the coupling parameter c for the solution landscapes. For simplicity, we fix the scaling $\xi = 1$ throughout this chapter.

For the visualisation, we plot the (headless) director $\mathbf{n}$ with rods and the normalised magnetisation vector field $\mathbf{m} = \frac{\mathbf{M}}{|\mathbf{M}|}$ with arrows.

7.1 Solver Details

The nonlinear solver is deemed to have converged when the Euclidean norm of the residual falls below 10^{-8}, or reduces from its initial value by a factor of 10^{-6}, whichever comes first. For the inner solver, the linearised systems are solved using the sparse LU factorisation library MUMPS (Amestoy *et al.*, 2000). We partition the whole interval $[-1, 1]$ into 1000 equi-distant subintervals and numerically approximate the solutions using $\mathbb{P}^1$ finite elements.

Code availability. For reproducibility and more details of the implementation, we have archived the solver code (Xia, 2021a) and the exact version of Firedrake (2021a) used to produce the numerical results of this work. An installation of Firedrake with components matching those used in this chapter can be obtained by following the instructions at https://www.firedrakeproject.org/download.html with

```
python3 firedrake-install --doi 10.5281/zenodo.4449535
```

Defcon version #aaa4ef should then be installed, as described in https://bitbucket.org/pefarrell/defcon/.

7.2   Solutions of the Full Problem

In this section, we focus on the full problem (5.10a)–(5.10d) with four scalar-valued solution variables $(Q_{11}, Q_{12}, M_1, M_2)$. We only present the result with small $k_1 = k_2 = k = 0.01$ (while varying the coupling c) here, since Theorem 5.4 implies that the OR solution branch remains as the unique minimiser of the full problem for a sufficiently large k and the OR solution will be reported later in Section 7.3. In fact, we shall see the uniqueness of solution for large k in the next section.

We first take the coupling parameter $c = 1$ and present four examples of stable stationary profiles $(Q_{11}, Q_{12}, M_1, M_2)$ in Figure 7.1. One can compare the L^∞ bound (5.11) with the computed values of vector norms $|\mathbf{Q}| = \sqrt{Q_{11}^2 + Q_{12}^2}$ and $|\mathbf{M}| = \sqrt{M_1^2 + M_2^2}$, and note that the pointwise maximum principle given by Theorem 5.2 is respected. By Remark 5.4, we expect that there is no interior domain wall with $|\mathbf{Q}| = |\mathbf{M}| = 0$, for small k, which is indeed noticeable from the presented solution profiles. Moreover, we can see that Solutions 1, 2 and 3 in Figure 7.1 only have boundary layers with constant $|\mathbf{Q}|, |\mathbf{M}|$-profiles in the interior domain, whereas Solution 4 has an interior non-zero local minimum (thus an interior jump) in $|\mathbf{Q}|$ and $|\mathbf{M}|$. In addition, Solutions 1 and 2 only differ in their orientational **m**-patterns (more precisely, possessing opposite signs of M_2) and they are the energy minimisers having the same energy value, while Solutions 3 and 4 are non-minimising stable critical points of the full energy (5.6).

Moreover, we compute the values of orientational angles θ and ϕ, defined as

$$\theta = \arctan\left(\frac{Q_{12}}{Q_{11}}\right), \quad \phi = \arctan\left(\frac{M_2}{M_1}\right) \tag{7.1}$$

for each numerical solution profile $(Q_{11}, Q_{12}, M_1, M_2)$, so to verify the relation (5.18), in particular the constraint (5.18c). It can be seen from Figure 7.1 that $|\mathbf{Q}| \to \rho^*, |\mathbf{M}| \to 1 + 2c\rho^*$ for the energy minimisers (Solutions 1 and 2), whereas $(2\phi - \theta)$ tends to be an even multiple of π almost everywhere, except close to the end point $y = 1$. Furthermore, we plot the separate values of θ and ϕ to demonstrate the linear behaviour consistent with (5.18) for ϕ and thus θ as $k \to 0$. This linearity of θ and ϕ can be seen in Figure 7.1 except around the local minima and boundary layers.

Now, we repeat the simulations for $c = 5$. Two stable stationary profiles are illustrated in Figure 7.2. Again, we observe that $|\mathbf{Q}| \to \rho*$ and

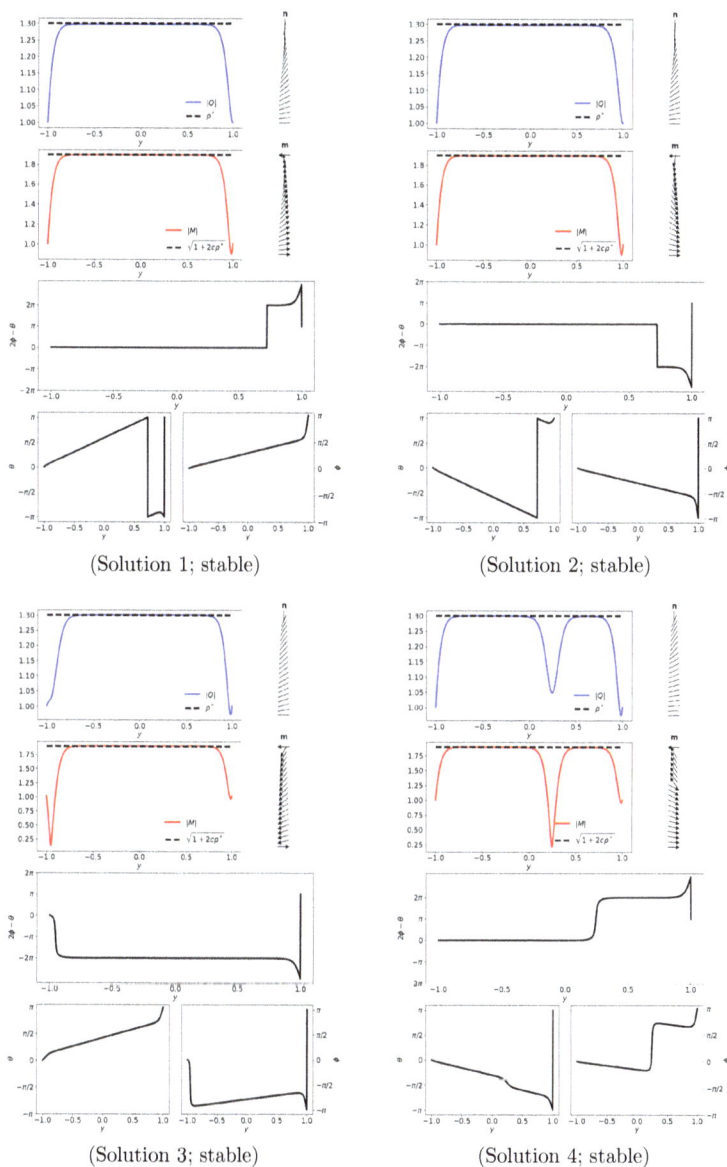

(Solution 1; stable)   (Solution 2; stable)

(Solution 3; stable)   (Solution 4; stable)

Fig. 7.1. Four stable stationary profiles, $(Q_{11}, Q_{12}, M_1, M_2)$, of (5.6) with $k = 0.01$ and $c = \xi = 1$, along with plots of $(2\phi - \theta)$, θ, and ϕ to verify the relation (5.18). Solutions 1 and 2 have the lowest full energy value (5.6).

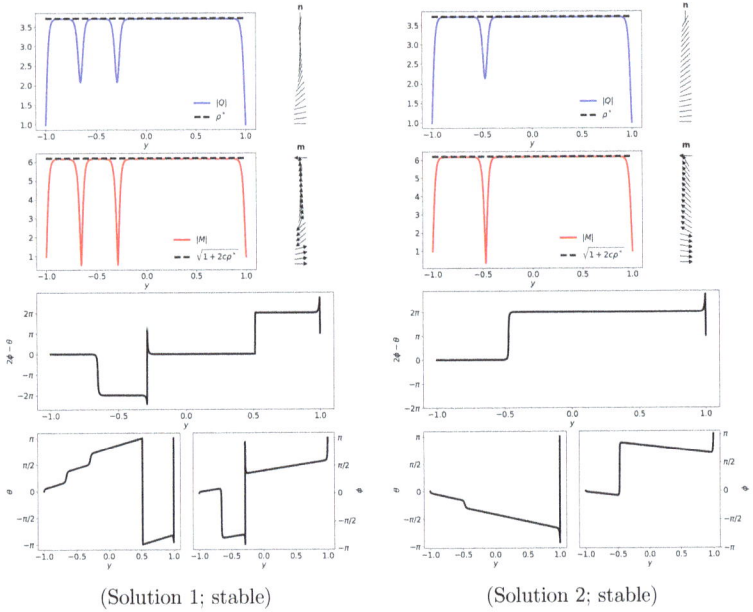

(Solution 1; stable)   (Solution 2; stable)

Fig. 7.2. Two examples of stable stationary profiles $(Q_{11}, Q_{12}, M_1, M_2)$ of the full energy (5.6) with $k = 0.01$, $c = 5$ and $\xi = 1$, along with plots of $(2\phi - \theta)$, θ, and ϕ to verify the relation (5.18). Solution 2 has lower energy than Solution 1.

$|\mathbf{M}|^2 \to 1 + 2c\rho^*$ almost everywhere, as expected from the maximum principle Theorem 5.2. Here, Solution 2 has lower energy than Solution 1, since Solution 1 has more jumps in $|\mathbf{Q}|$ and $|\mathbf{M}|$ than Solution 2. Further, $(2\phi - \theta)$ is an even multiple of π almost everywhere, with the jumps being associated with the jumps in $|\mathbf{Q}|$ and $|\mathbf{M}|$, thus verifying the constraint (5.18c). Additionally, we plot ϕ and θ in Figure 7.2, and observe almost linear profiles of θ and ϕ, except around the local minima and the boundary layers.

To summarise, the numerical experiments in this section and the theoretical heuristics in (5.18) suggest that there are at least two energy minimisers, characterised by $(\rho_1, \sigma_1, \theta_1, \phi_1)$ and $(\rho_2, \sigma_2, \theta_2, \phi_2)$ of the full ferronematic energy (5.6) in the $k \to 0$ limit, such that $\rho_1, \rho_2 \to \rho^*$, $\sigma_1^2, \sigma_2^2 \to 1 + 2c\rho^*$ almost everywhere away from the boundary plates $y = \pm 1$. Moreover, it holds that $\theta_2 = -\theta_1$, $\phi_2 = -\phi_1$ and $2\phi_{1,2} - \theta_{1,2}$ an even multiple of π except near $y = 1$ or close to some local jumps of $\mathbf{Q}$ and $\mathbf{M}$. The two energy minimisers only differ in the sense of rotation, in $\mathbf{n}$ and $\mathbf{m}$, between $y = -1$ and $y = 1$.

7.3 Solutions of the Reduced Problem

By the definition of the OR solution in Section 5.3, we know it is fully characterised by two degrees of freedom (Q_{11}, M_1) of the boundary-value problem (5.22) while $Q_{12} = M_2 = 0$ always holds. We now numerically investigate the limiting behaviours of the OR solution for $k \to 0$ and $k \to \infty$ illustrated in Section 5.3.

As $k \to \infty$, recall Theorem 5.1 to deduce that the OR solution branch is approximately given by $(\mathbf{Q}^{OR}, \mathbf{M}^{OR}) \approx (-y, 0, -y, 0)$, for a fixed c, and that $(\mathbf{Q}^{OR}, \mathbf{M}^{OR})$ is the unique minimiser of both the OR energy (5.19) and the full energy (5.6). In Figure 7.3, we plot the OR solution of (5.22) for $c = 1$ and $k = 10$. The profile is indeed linear, and we do not numerically obtain any other solutions, supporting the uniqueness result in this regime. We notice that the OR solution vanishes at the channel centre $y = 0$, i.e., $Q_{11}(0) = M_1(0) = 0$, and thus both the nematic and magnetic domain walls coincide at $y = 0$. Therefore, the normalised magnetisation vector $\mathbf{m}$ and director $\mathbf{n}$ have a jump discontinuity at $y = 0$. In fact, $\mathbf{m}$ jumps from $\mathbf{m} = (1, 0)$ for $y < 0$ to $\mathbf{m} = (-1, 0)$ for $y > 0$, while $\mathbf{n}$ jumps from $\mathbf{n} = (1, 0)$ (modulo a sign) for $y < 0$ to $\mathbf{n} = (0, 1)$ (modulo a sign) for $y > 0$. Hence, the nematic and magnetic domain walls at $y = 0$ separate two distinct polydomains in $\mathbf{n}$ and $\mathbf{m}$, respectively. We also plot the pointwise L^∞ bound (5.24) as blue solid lines in Figure 7.3, and as expected, this bound is indeed respected.

As $k \to 0$ with fixed positive c, the OR solution is not unique anymore and we expect to see that $(Q_{11}, M_1^2) \to (\rho^*, 1 + 2c\rho^*)$ uniformly everywhere away from the edges $y = \pm 1$, for the minimiser of the OR energy (5.19). Of course, all OR solutions are unstable critical points of the full energy (5.6) in the $k \to 0$ limit, as shown in Theorem 5.5. We now numerically corroborate these theoretical results with fixed $k = 0.01$ and $\xi = 1$.

In Figure 7.4, we present four example solutions by taking $c = 1$. In fact, they are all unstable critical points of the full energy (5.6) whilst being stable critical points of the OR energy (5.19) (in the sense that the Hessian of second variation of the OR energy about these critical points has only positive eigenvalues). Consistent with the discussion of the convergence regime for $k \to 0$ in Section 5.3, these solution profiles (Q_{11}, M_1) have a domain wall in $\mathbf{Q}$ near the end point $y = 1$, where Q_{11} jumps from $Q_{11} = \rho^* > 1$ to the boundary value $Q_{11}(1) = -1$. Analogously, we can see that all solution profiles illustrated in Figure 7.4 have a boundary layer close

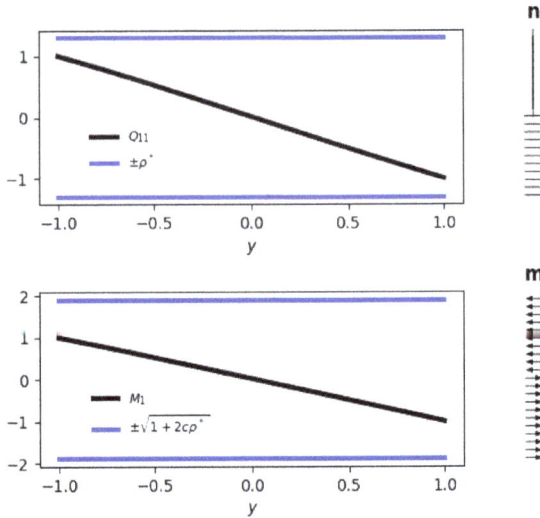

Fig. 7.3. The only (stable) solution of (5.19) for $c = \xi = 1$, and $k = 10$.

the other end point $y = -1$, within which Q_{11} jumps from $Q_{11}(-1) = 1$ to $Q_{11} = \rho^* > 1$. However, we should note that this boundary layer does not contain a domain wall with $Q_{11} = 0$. An additional observation is the presence of interior transition layers in M_1 (near the center $y = 0$) in Solutions 3 and 4 of Figure 7.4. The L^∞ bounds (5.24) (blue solid line) for $|Q_{11}|$ and $|M_1|$ are also satisfied.

In Figure 7.5, we plot the stable stationary profiles of the OR energy (5.19) for a larger value $c = 5$, whereas they are unstable critical points of the full energy (5.6). Indeed, each of the solutions in Figure 7.5 has one unstable eigendirection, in the context of the full energy (5.6). The two profiles in Figure 7.5, have boundary layers near $y = \pm 1$, and essentially differ in the sign of M_1 in the interior; Q_{11} only vanishes near $y = 1$, so that we have a nematic domain wall close to the end point $y = 1$. On the other hand, M_1 can vanish either near $y = -1$ or near $y = 1$, so that the corresponding magnetic domain wall can occur near either boundary. We also note that Solution 2 in Figure 7.5 is the OR energy minimiser which indeed converges to $\mathbf{p}^*$ almost everywhere except close to the boundary plates $y = \pm 1$. This verifies the heuristics explained by computing the transition costs in Figure 5.1.

Additionally, we present two more solution examples with interior transition layers for M_1 in Figure 7.6 with $c = 5$, where single and multiple

(Solution 1)   (Solution 2)

(Solution 3)   (Solution 4)

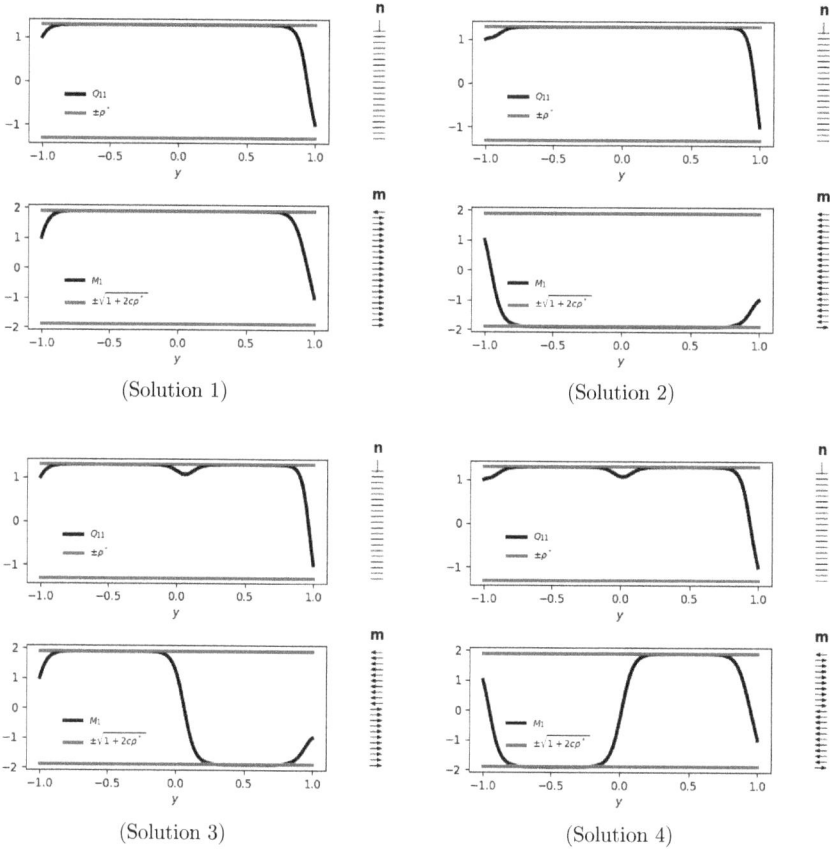

Fig. 7.4. Four OR solution profiles with $c = \xi = 1$ and $k = 0.01$. Solution 1 is the OR energy minimiser (5.19).

interior transition layers in M_1 are observed. They are also stable critical points of the OR energy (5.19), and unstable critical points of the full energy (5.6). The transition layers in M_1 necessarily contain a magnetic domain wall with $M_1 = 0$, and these interior magnetic domain walls are not accompanied by associated nematic domain walls. Moreover, solutions with interior transition layers have higher OR energy (5.19) than solutions without interior transition layers in Figure 7.5, since each transition layer requires an energetic cost of $\omega(p^*, p^{**})$. Again, the L^∞ bound (5.24) is satisfied for the solutions illustrated in Figure 7.6.

All above numerical experiments show that the domain walls in the OR energy minimisers migrate from the channel centre to the channel

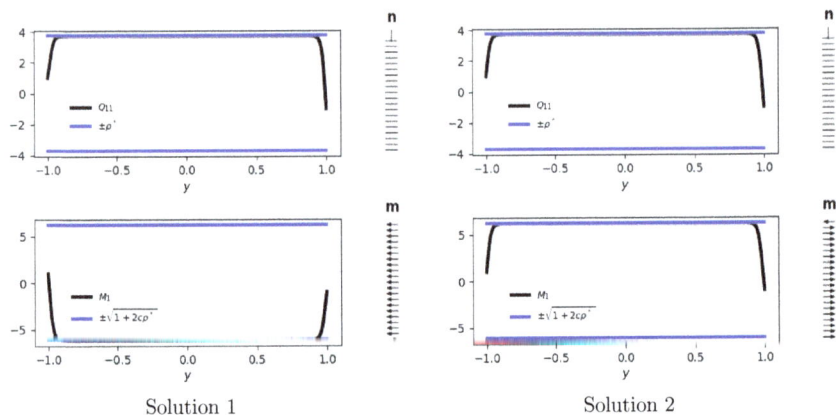

Solution 1   Solution 2

Fig. 7.5. Two stable OR critical points of (5.19), for $c = 5$, $\xi = 1$ and $k = 0.01$. The right profile has lower OR energy than the left profile and the solutions in Figure 7.6.

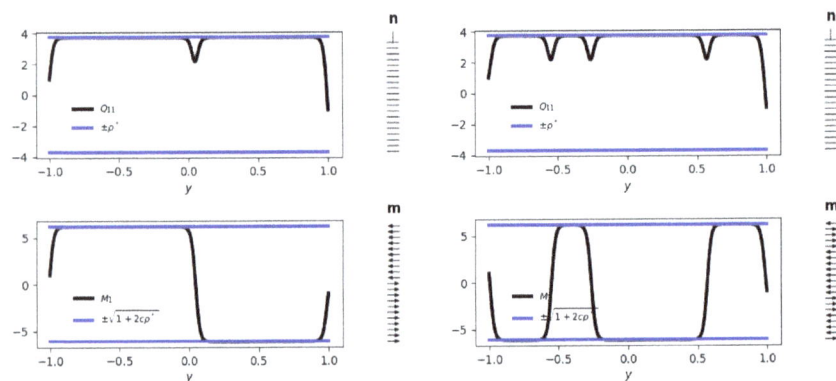

Fig. 7.6. Two stable OR solutions with single (left) and multiple (right) interior transition layers for $c = 5$, $\xi = 1$ and $k = 0.01$. The left profile has lower OR energy than the right profile.

boundaries at $y = \pm 1$, as k decreases. Therefore, we can manipulate the location and multiplicity of nematic and magnetic domain walls in the OR solutions by varying k.

7.4   Asymptotics Checking for $k \to \infty$

We then theoretically and numerically illustrate the asymptotic behaviour as $k \to \infty$ in this section, to investigate the convergence to the unique OR minimiser $(\mathbf{Q}^{OR}, \mathbf{M}^{OR}) = (-y, 0, -y, 0)$ in this limit regime.

As $k \to \infty$, we can compute useful asymptotic expansions of the OR solution branch for large k and small c, by setting $k = \frac{1}{c}$ in the Euler–Lagrange equations (5.10a)–(5.10d) and expanding around $(\mathbf{Q}^\infty, \mathbf{M}^\infty)$ as shown below:

$$Q_{11}(y) = -y + cf_2(y) + c^2 f_3(y) + \mathcal{O}(c^3),$$

$$M_1(y) = -y + cf_2^*(y) + c^2 f_3^*(y) + \mathcal{O}(c^3).$$

Substituting the above into (5.10a) and (5.10d) (with $k = \frac{1}{c}$) yields

$$\frac{d^2 f_1}{dy^2} + c\frac{d^2 f_2}{dy^2} + c^2\frac{d^2 f_3}{dy^2} = 4c\left(f_1^3 - f_1\right) + c^2\left(12f_1^2 f_2 - 4f_2 - (f_1^*)^2\right) + \mathcal{O}(c^3),$$

$$\tag{7.2a}$$

$$\frac{d^2 f_1^*}{dy^2} + c\frac{d^2 f_2^*}{dy^2} + c^2\frac{d^2 f_3^*}{dy^2} = c\left((f_1^*)^3 - f_1^*\right) + c^2\left(3(f_1^*)^2 f_2^* - f_2^* - \frac{2}{\xi}f_1 f_1^*\right)$$

$$+ \mathcal{O}(c^3). \tag{7.2b}$$

By equating powers of c, we solve the computed second-order ordinary differential equations for f_2, f_3, f_2^*, f_3^*, subject to the boundary conditions $f_2(-1) = f_2(1) = f_3(-1) = f_3(1) = 0$ and $f_2^*(-1) = f_2^*(1) = f_3^*(-1) = f_3^*(1) = 0$. This gives

$$c^0 : \frac{d^2 f_1}{dy^2} = 0 \Rightarrow f_1(y) = -y$$

$$c^1 : \frac{d^2 f_2}{dy^2} = 4(f_1^2 - 1)f_1 \Rightarrow f_2(y) = -\frac{1}{5}y^5 + \frac{2}{3}y^3 - \frac{7}{15}y$$

$$c^2 : \frac{d^2 f_3}{dy^2} = 4(3f_1^2 - 1)f_2 - (f_1^*)^2 \Rightarrow f_3(y) = p(y),$$

and

$$c^0 : \frac{d^2 (f_1^*)}{dy^2} = 0 \Rightarrow f_1^*(y) = -y,$$

$$c^1 : \frac{d^2 (f_2^*)}{dy^2} = ((f_1^*)^2 - 1)f_1^* \Rightarrow f_2^*(y) = -\frac{1}{20}y^5 + \frac{1}{6}y^3 - \frac{7}{60}y,$$

$$c^2 : \frac{d^2 (f_3^*)}{dy^2} = 3(f_1^*)^2 f_2^* - f_2^* - \frac{2}{\xi}f_1 f_1^* \Rightarrow f_3^*(y) = q(y).$$

Here,

$$p(y) = -\frac{1}{30}y^9 + \frac{22}{105}y^7 - \frac{31}{75}y^5 - \frac{1}{12}y^4 + \frac{14}{45}y^3 - \frac{233}{3150}y + \frac{1}{12},$$

and

$$q(y) = -\frac{1}{480}y^9 + \frac{11}{840}y^7 - \frac{31}{1200}y^5 - \frac{1}{6}y^4 + \frac{7}{360}y^3 - \frac{233}{50400}y + \frac{1}{6}.$$

Thus, the expansions for Q_{11} and M_1 are

$$Q_{11}(y) = -y + c\left(-\frac{1}{5}y^5 + \frac{2}{3}y^3 - \frac{7}{15}y\right) + c^2 p(y) + \mathcal{O}(c^3), \qquad (7.3)$$

and

$$M_1(y) = -y + c\left(-\frac{1}{20}y^5 + \frac{1}{6}y^3 - \frac{7}{60}y\right) + c^2 q(y) + \mathcal{O}(c^3), \qquad (7.4)$$

for $k = \frac{1}{c}$ and $k \gg 1$.

We now check the validity of these expansions, (7.3) and (7.4), numerically. To this end, we compare $(\cdot)^{\mathrm{num}}$ and $(\cdot)^{\mathrm{asymp}}$ in the L^∞-norm, where $(\cdot)^{\mathrm{num}}$ is the numerical solution and $(\cdot)^{\mathrm{asymp}}$ corresponds to the asymptotic expansion, depending on the truncation of the expansions in (7.3) and (7.4). For instance, a first-order truncation (with respect to c) yields

$$Q_{11}^{\mathrm{asymp}} = -y + c\left(-\frac{1}{5}y^5 + \frac{2}{3}y^3 - \frac{7}{15}y\right),$$

$$M_1^{\mathrm{asymp}} = -y + c\left(-\frac{1}{20}y^5 + \frac{1}{6}y^3 - \frac{7}{60}y\right).$$

The left-hand column of Figure 7.7 shows a first-order convergence by truncating the expansions up to $\mathcal{O}(c^0)$, whilst a first-order truncation leads to a second-order convergence as shown in the middle column of Figure 7.7 and finally, in the right-hand column, a truncation up to $\mathcal{O}(c^2)$ demonstrates a third order convergence with respect to c, for both Q_{11} and M_1.

7.5   Bifurcation Diagrams

The proceeding sections examine the behaviour of the solution profiles for certain specific choices of parameters. One can obtain further information about the solutions to the Euler–Lagrange equations (5.10a)–(5.10d) by continuing the parameter and plotting bifurcation diagrams for the parameter space of interest. We thus perform numerical experiments as we continue the coupling parameter c or the elastic constant k.

The first experiment regarding varying c is illustated in Figure 7.8. Here, we choose $k_1 = k_2 = k \in [0.2, 3.0]$ with fixed step size 0.01 and $c = 1$.

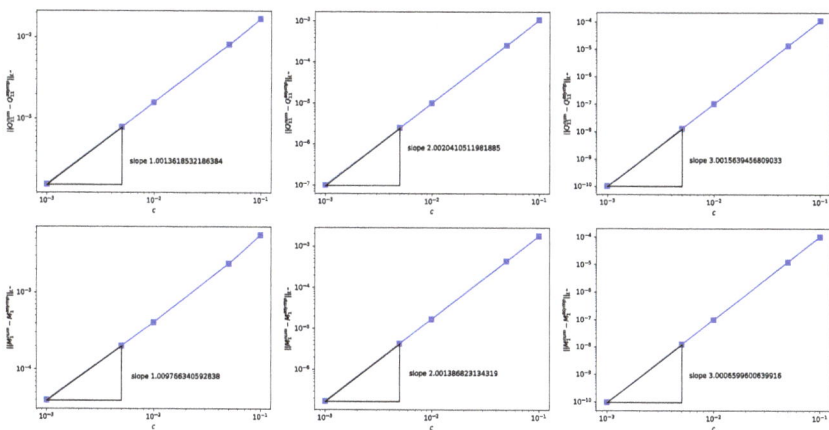

Fig. 7.7. Log–log plots of $\|Q_{11}^{\text{num}} - Q_{11}^{\text{asymp}}\|_\infty$ (top row) and $\|M_1^{\text{num}} - M_1^{\text{asymp}}\|_\infty$ (bottom row). Left: truncating asymptotic expansions (7.3) and (7.4) at c^0. Middle: truncating asymptotic expansions at c^1. Right: truncating asymptotic expansions at c^2.

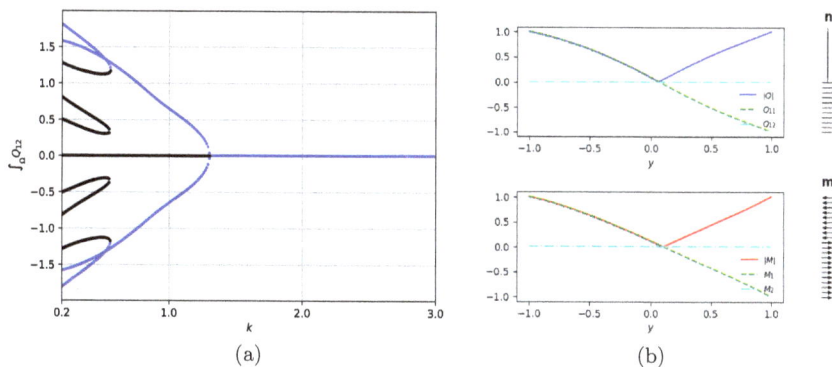

(a)   (b)

Fig. 7.8. (a) The bifurcation diagram of continuing $k_1 = k_2 = k \in [0.2, 3.0]$ with fixed $c = \xi = 1$; here, black represents unstable solutions while blue indicates stable solutions. (b) The stable solution for $k = 2$.

It can be seen that there is only one stable OR solution for $k \in [1.25, 3.0]$, being the energy minimiser of the full energy (5.6). For $k \approx 1.25$, there is a pitchfork bifurcation consisting of two stable branches and one unstable OR branch (see Figure 7.9 for an illustration of these three solutions at $k = 1$). In fact, the two stable solutions (Solutions 1 and 3 in Figure 7.9) differ by the sign of Q_{12} and M_2, i.e., for every solution branch, $(Q_{11}, Q_{12}, M_1, M_2)$,

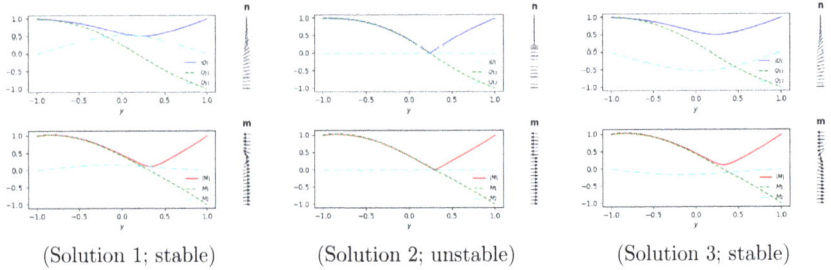

<table>
<tr><td>(Solution 1; stable)</td><td>(Solution 2; unstable)</td><td>(Solution 3; stable)</td></tr>
</table>

Fig. 7.9. Three solutions for $k = 1$ in Figure 7.8. Solutions 1 and 3 are global energy minimisers.

there exists another solution branch with $(Q_{11}, -Q_{12}, M_1, -M_2)$. The stable solution branches correspond to a smooth rotation in $\mathbf{n}$, between the two end points $y = \pm 1$ and are actually the global energy minimisers for $k \leq 1.25$.

As k becomes smaller, more (stable or unstable) solutions are found. More specifically, there are four disconnected bifurcations appearing around $k = 0.55$, giving two further stable solutions, which are also local energy minimisers (see Solutions 1 and 8 in Figure 7.10 for an illustration) for $k \in [0.2, 0.55]$. Again, they only differ by the sign of Q_{12} and M_2. In Figure 7.10, we plot eight newly found solution profiles, along with their stabilities. The stable solutions typically correspond to a smooth $\mathbf{n}$-profiles with minimal rotation (minimal topological degree consistent with the boundary conditions), while the stable normalised magnetisation profiles $\mathbf{m}$ are also smooth, except for a thin interval of large rotation in $\mathbf{m}$ localised near the end points $y = \pm 1$. Meanwhile, it can be seen that the unstable solution pairs, i.e., Solutions 2 and 7, Solutions 3 and 6 and Solutions 4 and 5 also differ by the sign of Q_{12} and M_2. Interestingly, all profiles with interior jumps in $\mathbf{n}$ and $\mathbf{m}$ are unstable.

We next investigate the loss of stability of the OR solution branch for a larger value of c, i.e., we numerically compute a bifurcation diagram in Figure 7.11, for the solutions of (5.10a)–(5.10d), by continuing $k \in [3, 5]$ with a step size of 0.015, and fixed $c = 5$. One stable OR solution is shown in Figure 7.11 and it loses stability at the pitchfork bifurcation point $k \approx 4.46$, leading to two new stable branches (see illustrations in Figure 7.12 for $k = 4.43$). We observe that they only differ in the signs of Q_{12} and M_2 and in fact are energy minimisers for $k \leq 4.43$. Thus, the qualitative features of the bifurcation diagram are unchanged by increasing c but the OR solution

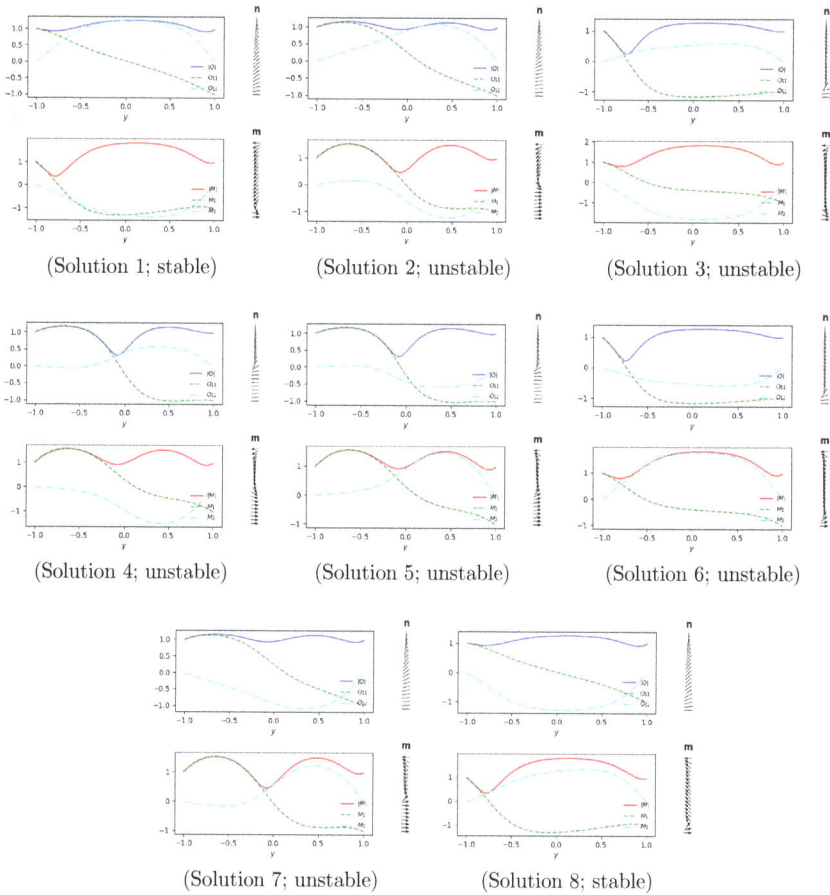

(Solution 1; stable)   (Solution 2; unstable)   (Solution 3; unstable)

(Solution 4; unstable)   (Solution 5; unstable)   (Solution 6; unstable)

(Solution 7; unstable)   (Solution 8; stable)

Fig. 7.10. Eight new solutions for $k = 0.2$ in Figure 7.8. Solutions 1 and 8 are global energy minimisers.

branch loses stability for $k < k^*(c)$, where $k^*(c)$ is an increasing function of c. Hence, as c increases, OR solutions are increasingly difficult to find owing to their shrinking window of stability.

Remark 7.1. One may wonder about the appearance of the two folds in the bifurcation diagram depicted in Figure 7.11. They do not represent the same solution at the intersection points. Instead, they are just overlapping points in this plot of $\int_\Omega Q_{12}$ against k. Choosing a different functional may yield a bifurcation diagram without these intersection points.

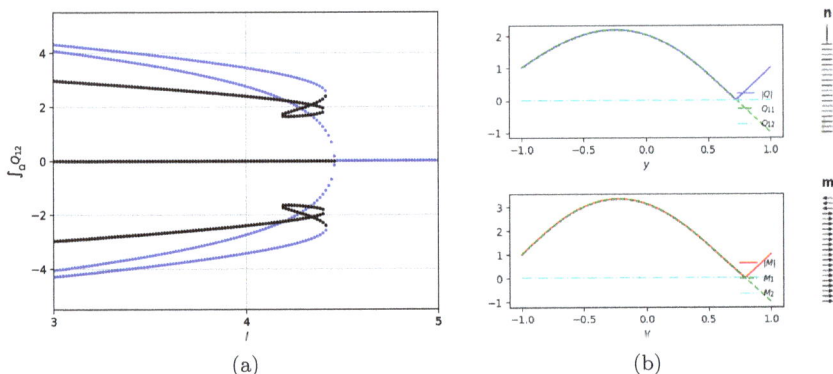

(a)   (b)

Fig. 7.11.   (a) The bifurcation diagram with fixed $c = 5$ and $\xi = 1$; here, black labels unstable solutions while blue labels stable solutions. (b) One stable OR solution for $k = 4.45$.

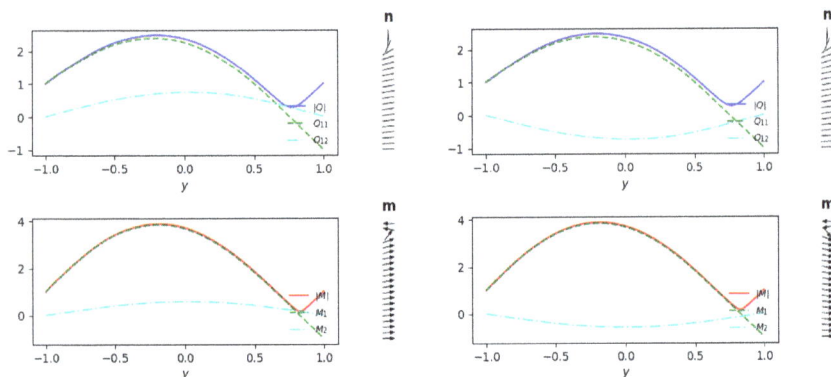

Fig. 7.12.   Two new stable solutions at $k = 4.43$ in Figure 7.11.

7.6   Summary

In this chapter, we performed several numerical experiments that validate the theoretical analysis derived in Dalby *et al.* (2022). These include providing more complete solution landscapes of the ferronematic problem, stability analysis, and showing multiple patterns of domain walls in the interior. We demonstrated the strength of **Q**-tensor theory for characterising defects (i.e., domain walls in director **n** and normalised magnetisation **m**) in one-dimensional ferronematics. We will further consider more complicated defect structures (in higher dimensions) in the next part of this book.

Part 3
Smectic Liquid Crystals

Chapter 8

A Mathematical Model of Smectics*

In the proceeding part, we have considered the application of the $\mathbf{Q}$-tensor theory in ferronematics, which can possess multiple domain walls (i.e., where the nematic tensor $\mathbf{Q}$ or magnetic order parameter $\mathbf{M}$ vanishes) separating polydomains. In this part, we study and model more complicated defect structures that exist in smectics, more precisely, in the smectic-A phase.

Smectic liquid crystals are layered mesophases that have a periodic modulation of the mass density along one spatial direction. Roughly speaking, they can be thought of as one-dimensional solids along the direction of periodicity and two-dimensional fluids along the other two remaining directions. Due to their periodic structures, smectic liquid crystals have drawn extensive research attention and are directly related to some applications in photonic band-gap materials, metamaterials, and templates for guided particle self-assembly (Zappone and Lacaze, 2008).

Two common phases of smectic liquid crystals are the smectic-A and smectic-C phases (see Figure 8.1 for an illustration). In smectic-A phases, the director is parallel to the normal of the smectic layers while smectic-C phases allow the director to freely rotate around the normal, and thus present a tilted angle between the director and the layer normals. In order to characterise the periodic property of the density in smectic phases, de Gennes first proposed to use a complex-valued variable as the smectic order parameter, based on an analogue to superfluids in superconductors (de Gennes, 1972). This theory (abbreviated as the dG theory) for modelling smectics has been a popular tool for investigating defect structures in

---

*This part of work majorly expands on the author's work in Xia *et al.* (2021b) and Xia and Farrell (2023).

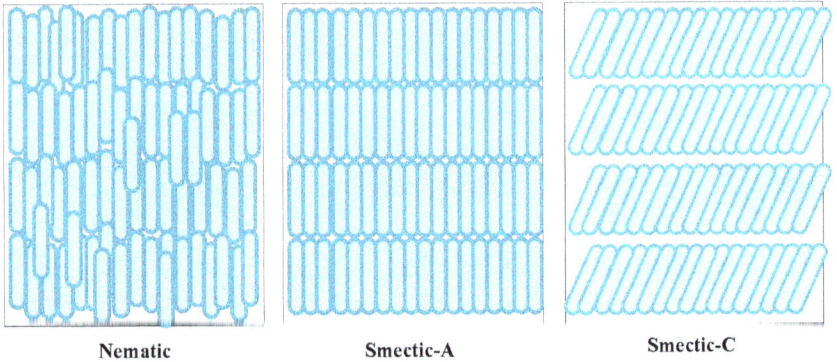

| Nematic | Smectic-A | Smectic-C |

Fig. 8.1. Graphical illustrations of nematic, smectic-A and smectic-C phases. The top and bottom substrate plates are polarisers with perpendicular alignment directions. This type of polarisers is for example used in twisted nematic display (Dunmur and Sluckin, 2011, Technical Box 10.1).

smectic phases, e.g., Ogawa and Uchida (2006) and Santangelo and Kamien (2007) and for modelling smectic liquid crystal fluids (E, 1997).

In this chapter, we first review the classical dG model for smectic-A liquid crystals and then a more recent model by Pevnyi *et al.* (2014) using a real-valued smectic order parameter. Next, we propose a new model inheriting the advantages of the real-valued smectic model, which can also represent half charge defects by adopting a **Q**-tensor as the nematic order parameter.

8.1   The de Gennes Model

According to de Gennes' theory (de Gennes, 1972, 1974), one can model smectic liquid crystals based on a complex-valued order parameter $\psi : \Omega \to \mathbb{C}$, which describes the magnitude $|\psi|$ and the phase $\nabla \psi$ of smectic layer ordering, and a real vector-valued nematic order parameter **n** satisfying the unit-length constraint $|\mathbf{n}|=1$. Furthermore, the phase $\nabla \psi$ indicates the position of the layers. There is a strong analogy between the derivation of de Gennes's formulation for smectics and that of superconductors, as discussed in de Gennes (1972) and Halperin and Lubensky (1974).

More precisely, de Gennes proposed the free energy of smectic-A LC to be

$$\mathcal{J}^{\mathrm{dG}}(\mathbf{n}, \psi) = \int_{\Omega} \big(F_S(\mathbf{n}, \psi) + W^{OF}(\mathbf{n}) \big), \qquad (8.1)$$

where $\Omega \subset \mathbb{R}^d$ ($d \in \{2,3\}$) is the region occupied by liquid crystals, W^{OF} denotes the nematic Oseen–Frank energy density of the form (2.3), and F_S represents the smectic energy density given by

$$F_S(\mathbf{n}, \psi) = |\nabla \psi - iq\mathbf{n}\psi|^2 + \varsigma|\psi|^2 + \frac{\varpi}{2}|\psi|^4. \qquad (8.2)$$

Here, $i = \sqrt{-1}$, q represents the length of the favoured wave-vector, $\varpi > 0$ a fixed number and $\varsigma = \varsigma_0(T_m - T_{ns})$ the discrepancy between the material temperature T_m and nematic-smectic transition temperature T_{ns}, with $\varsigma_0 > 0$. Since we are focusing on the smectic phase where T_m is normally below the transition temperature T_{ns}, it holds that $\varsigma < 0$.

It is obvious to see that when $\psi = 0$, (8.1) reduces to the nematic phase; and $\psi \neq 0$ corresponds to the smectic phase. Furthermore, one can note there is no odd power of the amplitude $|\psi|$ in the smectic energy density (8.2). This is because a change in sign, $\psi \to -\psi$, corresponds to a uniform translation of the smectic layers by one smectic layer and it should cost no additional energy to do so (Linhananta and Sullivan, 1991).

Remark 8.1. We have some comments regarding the derivation of dG model (8.1) for smectic liquid crystals: (a) the coefficients in (8.1) are phenomenological and their relations to molecular properties are not revealed (Linhananta and Sullivan, 1991); (b) the smectic order parameter ψ is assumed to vary spatially on a length scale larger than the layer thickness τ; (c) the free energy (8.1) only includes independent fluctuations (i.e., the W^{OF} density term and the ς, ϖ term in F_S are dependent only on $\mathbf{n}$ and ψ respectively) in the quantities ψ and $\mathbf{n}$; (d) no orientational order parameter (e.g., tensor order parameter $\mathbf{Q}$) has been involved. Linhananta and Sullivan (1991) have presented a modified dG energy to overcome the above limitations by means of molecular density functional theories.

It is important to understand what the coupling term $|\nabla\psi - iq\mathbf{n}\psi|^2$ describes. To this end, we can express the smectic order parameter by

$$\psi(\mathbf{x}) = \varrho(\mathbf{x})e^{i\iota(\mathbf{x})}, \quad \varrho : \Omega \to \mathbb{R}, \quad \iota : \Omega \to \mathbb{R}, \qquad (8.3)$$

where $\varrho(\mathbf{x}) = |\psi(\mathbf{x})|$ denotes the mass density of the smectic layers at a point $\mathbf{x} \in \Omega$ and ι parametrises the layers so that $\nabla\iota$ indicates the direction of the layer normal. Substituting the above expression into the coupling term, we obtain

$$|\nabla\psi - iq\mathbf{n}\psi|^2 = |\nabla\varrho|^2 + \varrho^2|\nabla\iota - q\mathbf{n}|^2,$$

and the smectic energy density F_S becomes

$$F_S(\mathbf{n}, \varrho, \iota) = |\nabla \varrho|^2 + \varrho^2|\nabla \iota - q\mathbf{n}|^2 + \varsigma|\varrho|^2 + \frac{\varpi}{2}|\varrho|^4. \qquad (8.4)$$

Consequently, as we perform minimisation over F_S, we are actually penalising the nematic-smectic coupling constraint $\nabla \iota = q\mathbf{n}$. This illustrates how smectic layers align with nematic directors $\mathbf{n}$, that is to say, the smectic layer normals should be parallel to the director.

Remark 8.2. If $\mathbf{n}$ is a gradient (i.e., $q\mathbf{n} = \nabla \iota$ which can be derived from penalising the coupling term $\varrho^2|\nabla \iota - q\mathbf{n}|^2$ in (8.4)), then the twist-effect $\mathbf{n} \cdot (\nabla \times \mathbf{n})$ in $W^{OF}(\mathbf{n})$ is zero. This is known as the incompatibility between smectic order and twist (see e.g., Calderer and Palffy-Muhoray, 2000, Section 1.6; Santangelo and Kamien, 2007).

Moreover, the molecular mass density is defined as

$$\varrho_m(\mathbf{x}) = \varrho_0 + \frac{1}{2}(\psi(\mathbf{x}) + \psi^*(\mathbf{x})) = \varrho_0 + \varrho(\mathbf{x})\cos\iota(\mathbf{x}) = \varrho_0 + |\psi(\mathbf{x})|\cos\iota(\mathbf{x}),$$
$$(8.5)$$

where ϱ_0 is the average density and ψ^* represents the complex conjugate of ψ. Hence, $|\psi(\mathbf{x})|\cos\iota(\mathbf{x})$ gives the real-valued density variation between the molecular mass density and the average density. The derivation of the model by Pevnyi *et al.* (2014) to be introduced in the next section in fact utilises such a real variable as the smectic order parameter, as we shall now see.

8.2   The Pevnyi–Selinger–Sluckin Model

As discussed in Bedford (2014) and Pevnyi *et al.* (2014), the classical dG model (8.1) using the complex order parameter ψ gives rise to a direct difficulty: $\text{Im}(\psi)$ does not relate to anything physical. A resulting branch-cut due to the presence of this issue is schematically illustrated in Pevnyi *et al.* (2014, Fig. 1) with a $+1/2$-charge disclination. This situation is similar to the case of representing the $+1/2$-charge defect by the vector-valued director $\mathbf{n}$, where the head-to-tail symmetry of molecules is not respected and thus a branch-cut occurs when $\mathbf{n}$ changes to $-\mathbf{n}$. To avoid the use of a complex variable, Pevnyi *et al.* (2014) proposed a new model (abbreviated henceforth as the *PSS* model) adopting the director $\mathbf{n} : \Omega \to \mathbb{R}^d$ and the density variation $u : \Omega \to \mathbb{R}$ from the average density as state variables.

The form of the PSS free energy is given by

$$\mathcal{J}^{PSS}(u, \mathbf{n}) = \int_\Omega \left(f_s(u) + \frac{K}{2}|\nabla \mathbf{n}|^2 + B \left|\mathcal{D}^2 u + q^2 (\mathbf{n} \otimes \mathbf{n}) u\right|^2 \right), \quad (8.6)$$

where the smectic bulk energy density is given by

$$f_s(u) = \frac{a_1}{2}u^2 + \frac{a_2}{3}u^3 + \frac{a_3}{4}u^4.$$

Here a_1, a_2, a_3, B, K and q are some known real parameters. Moreover, the unit length constraint $\mathbf{n} \cdot \mathbf{n} = 1$ for the director must be enforced. In order to keep f_s bounded from below, we need to choose $a_3 > 0$, and to possess nonzero (i.e., $u \neq 0$) minimisers of f_s (thus not pure nematic minimisers), we should choose $a_1 < 0$.

Remark 8.3. One can notice that a cubic term of u is added to f_s in (8.6) when comparing it with the dG model (8.1). This is allowed because we should not expect symmetry between positive density variation $u > 0$ and negative density variation $u < 0$.

The derivation of the PSS model comes from the density functional theory (based on a molecular statistical description) analogous to early work in Linhananta and Sullivan (1991) and Poniewierski and Sluckin (1991); however, a detailed explanation (in particular, about how the model parameters are related to some physically measurable constants) is not given in Pevnyi *et al.* (2014). In fact, the idea of Linhananta and Sullivan (1991) is to divide the total free energy into local and nonlocal parts. The local energy includes an isotropic term, modelled by the standard quartic order Landau–Ginzburg free energy with regard to the smectic density variable, and an anisotropic term of $\mathbf{Q}$-tensor to characterise nematic LC. For the nonlocal part, they adopt the typical form of two-body contributions to the free energy occurring in mean-field density functional theories, which gives rise to a fourth order term similar to the coupling B-term in (8.6).

For a better understanding of the PSS model, particularly how the coupling term in (8.6) relates to the physical constraint of smectics, we give our interpretation in the following. As described in Pevnyi *et al.* (2014) and illustrated in (8.5), the density variation u can be related to the complex order parameter ψ in the dG model by the expression

$$u = \Re\psi = |\psi| \cos(\iota)$$

with $\Re$ denotes the real part of a complex number. Note that the amplitude $|\psi|$ of the density modulation does not vary spatially as it refers to the largest mass density and we can actually see this fact in the numerical results in Chapter 10. From what we have discussed in Section 8.1, minimising $|\nabla\psi - iq\mathbf{n}\psi|^2$ in fact promotes the relation $\nabla\iota = q\mathbf{n}$. Subsequently combining with (8.3), one can expect the following expression of ψ,

$$\psi(\mathbf{x}) = |\psi|e^{iq\mathbf{n}\cdot\mathbf{x}}.$$

Therefore, we obtain the corresponding form of the density variation u as follows:

$$u(\mathbf{x}) = |\psi|\cos(q\mathbf{n}\cdot\mathbf{x}). \tag{8.7}$$

We then calculate

$$\mathcal{D}^2u = \mathcal{D}(\mathcal{D}u) = \mathcal{D}(-|\psi|\sin(q\mathbf{n}\cdot\mathbf{x})q\mathbf{n}) = -q^2(\mathbf{n}\otimes\mathbf{n})u,$$

and it follows that

$$\mathcal{D}^2u + q^2(\mathbf{n}\otimes\mathbf{n})u = 0.$$

Hence, one can interpret minimising the coupling term $|\mathcal{D}^2u + q^2(\mathbf{n}\otimes\mathbf{n})u|^2$ as respecting the periodicity of the smectic density, i.e., $u = |\psi|\cos(q\mathbf{n}\cdot\mathbf{x})$.

The PSS model helps investigate defect structures appearing in the smectic-A phase in a more physically reasonable way without using the complex order parameter ψ in the classical dG model (8.1). There are some numerical examples of smectic layers respecting different topological defects illustrated in the work (Pevnyi *et al.*, 2014). However, by solving the PSS model as described using $\mathbf{n} \in H^1(\Omega, \mathcal{S}^{d-1})$, we cannot reproduce the experiments of half-charge defects that are shown in Pevnyi *et al.* (2014). This is due to the presence of a discontinuity in the director $\mathbf{n}$ in these defects, which cannot be characterised by a continuous vector field (Ball, 2017). As a matter of fact, in private communication, the authors of Pevnyi *et al.* (2014) have commented that they actually implemented their model with the tensor product $\mathbf{n}\otimes\mathbf{n}$, thus enforcing the unit length constraint of director $\mathbf{n}$ implicitly through introducing the tensor $\mathbf{n}\otimes\mathbf{n}$. This allows them to represent half-charge defects (Ball, 2017), but numerically enforcing that the order parameter is a line field of the form $\mathbf{n} \otimes \mathbf{n}$ in minimisation is difficult (Borthagaray *et al.*, 2020).

8.3 Our Proposed Model

A new mathematical model that incorporates both a tensor field and a real-valued density variation field could be useful in representing smectic liquid crystals with complex defect structures. In fact, the idea of combining a $\mathbf{Q}$-tensor variable and a real-valued density variable to model smectic LC has been previously discussed in Han *et al.* (2015), Linhananta and Sullivan (1991), and Mei and Zhang (2015). However, these works are all molecular-based microscopic models which are difficult to implement due to their natural complexity in relating statistical parameters to physically realistic experimental results. It is an open problem to combine both the microscopic and macroscopic sides for modelling smectic liquid crystals, as discussed by Ball and Bedford (2015). Moreover, these authors (Ball and Bedford, 2015 and Bedford, 2014) have noticed the necessity of combining the nematic order parameter $\mathbf{Q}$ and the real-valued smectic order parameter to characterise defects and thus modified the PSS model by replacing $\mathbf{n} \otimes \mathbf{n}$ by $(\mathbf{Q}/s + \mathbf{I}_3/3)$ arising from the uniaxial expression of $\mathbf{Q}$-tensor:

$$
\mathcal{J}^{BB}(u, \mathbf{Q}) = \int_\Omega \frac{K}{2} |\nabla \mathbf{Q}|^2 + B \left| \mathcal{D}^2 u + q^2 \left(\frac{\mathbf{Q}}{s} + \frac{\mathbf{I}_3}{3} \right) u \right|^2
$$
$$
+ \frac{a_1}{2} u^2 + \frac{a_2}{3} u^3 + \frac{a_3}{4} u^4, \tag{8.8}
$$

with $u \in H^2(\Omega, \mathbb{R})$ and $\mathbf{Q} \in SBV(\Omega, S_0)$ where SBV denotes special functions of bounded variation. A preliminary result of existence of minimisers for their modified model is also briefly included. Nevertheless, the possibility of characterising defects existing in smectic liquid crystals and the implementation of their model has not been investigated or realised. One can readily notice the numerical singularities caused by the denominator s whenever it is near zero (which is likely to happen around defects). To avoid the aforementioned issue of the denominator s, we assume that the scalar order parameter s is a fixed constant, which can be determined by the form of the additional nematic bulk energy (we will discuss this point in detail later) arising from the phenomenological LdG model of nematics.

8.3.1 *A unified framework*

In this part, we further assume that Ω is convex as such convexity is needed for the regularity result (see Theorem 9.1).

Considering that smectic-A liquid crystals are optically uniaxial (de Gennes, 1973, 1974), we can express the $\mathbf{Q}$ tensor in a uniaxial form: $\mathbf{Q} = s \left(\mathbf{n} \otimes \mathbf{n} - \frac{\mathbf{I}_d}{d} \right)$, where the director $\mathbf{n}$ is the corresponding eigenvector of $\mathbf{Q}$ with the major eigenvalue, say, λ_{eig}. One can readily check that s and

λ_{eig} satisfy the relation

$$s = 2\lambda_{\mathrm{eig}} \quad \text{for } d = 2,$$

$$s = \frac{3}{2}\lambda_{\mathrm{eig}} \quad \text{for } d = 3.$$

Moreover, the symmetric traceless $\mathbf{Q}$-tensor has two degrees of freedom (Q_{11}, Q_{12}) in two dimensions or five degrees of freedom $(Q_{11}, Q_{12}, Q_{13}, Q_{22}, Q_{23})$ in three dimensions. Thus, it can be expressed in the form of

$$\mathbf{Q} = \begin{bmatrix} Q_{11} & Q_{12} \\ Q_{12} & -Q_{11} \end{bmatrix} \quad \text{or} \quad \mathbf{Q} = \begin{bmatrix} Q_{11} & Q_{12} & Q_{13} \\ Q_{12} & Q_{22} & Q_{23} \\ Q_{13} & Q_{23} & -(Q_{11} + Q_{22}) \end{bmatrix}. \quad (8.9)$$

In particular, we note that $\mathrm{tr}(\mathbf{Q}^3) = 0$ for $d = 2$ which can be easily checked via computations using (8.9).

We now propose the following $\mathbf{Q}$-tensor model that incorporates the dG theory for smectic LC and LdG model for nematics while also keeping the density variable u to be real-valued, as discussed in Pevnyi *et al.* (2014):

$$\mathcal{J}(u, \mathbf{Q}) = \int_{\Omega} \left(f_s(u) + B \left| \mathcal{D}^2 u + q^2 \left(\mathbf{Q} + \frac{\mathbf{I}_d}{d} \right) u \right|^2 + f_n(\mathbf{Q}, \nabla\mathbf{Q}) \right), \quad (8.10)$$

where

$$f_s(u) := \frac{a_1}{2} u^2 + \frac{a_2}{3} u^3 + \frac{a_3}{4} u^4 \quad (8.11)$$

and

$$f_n(\mathbf{Q}, \nabla\mathbf{Q})$$
$$= f_n^e(\nabla\mathbf{Q}) + f_n^b(\mathbf{Q})$$
$$:= \frac{K}{2} |\nabla\mathbf{Q}|^2 + \begin{cases} \left(-l \left(\mathrm{tr}(\mathbf{Q}^2) \right) + l \left(\mathrm{tr}(\mathbf{Q}^2) \right)^2 \right), & \text{if } d = 2, \\ \left(-\frac{l}{2} \left(\mathrm{tr}(\mathbf{Q}^2) \right) - \frac{l}{3} \left(\mathrm{tr}(\mathbf{Q}^3) \right) + \frac{l}{2} \left(\mathrm{tr}(\mathbf{Q}^2) \right)^2 \right), & \text{if } d = 3. \end{cases}$$
$$(8.12)$$

Here, K is the nematic elastic constant, l represents the nematic bulk parameter that can depend on temperature and a_1, a_2, a_3, B, q are inherited from the PSS model. We refer to the *decoupled* case when $q = 0$.

Remark 8.4. We can observe some differences between our proposed model (8.10) and the Ball–Bedford model (8.8): (a) we have taken the scalar order parameter s to be a fixed constant (in fact, $s = 1$), which is weakly preferred due to the addition of the nematic bulk term $f_n^b(\mathbf{Q})$; (b) we give a unified framework applicable to both two and three dimensions.

One may notice that the term f_n arises in the classical LdG model (5.4) for nematic LC. Furthermore, it is known that the global minimiser of the bulk energy f_n^b is a uniaxial $\mathbf{Q}$ tensor with scalar order parameter $s = 1$ (one can check this by some calculations or using (Majumdar and Zarnescu, 2010, Proposition 15) as quoted below for self containment). Adding the bulk energy terms helps in deciding the scalar order parameter s, and therefore we can adjust the coefficients in the bulk energy density so to promote $s = 1$ almost everywhere.

Proposition 8.1 (Majumdar and Zarnescu, 2010, Proposition 15). *Assume that l_a, l_b, l_c are positive parameters and consider the bulk energy in the following form:*

$$f_n^b(\mathbf{Q}) = -\frac{l_a}{2}\left(tr(\mathbf{Q}^2)\right) - \frac{l_b}{3}\left(tr(\mathbf{Q}^3)\right) + \frac{l_c}{4}\left(tr(\mathbf{Q}^2)\right)^2. \qquad (8.13)$$

Then its minimiser is a uniaxial tensor of the form

$$\mathbf{Q} = s_+\left(\mathbf{n} \otimes \mathbf{n} - \frac{\mathbf{I}_3}{3}\right),$$

where

$$s_+ = \frac{l_b + \sqrt{l_b^2 + 24 l_a l_c}}{4 l_c}.$$

8.3.2 *Existence of minimisers*

We have proposed a unified functional (8.10) for both two- and three-dimensional cases to be minimised on some admissible set. An immediate question is whether minimisers exist.

We define the admissible space $\mathcal{A}^s$ of our proposed functional $\mathcal{J}$ as

$$\mathcal{A}^s = \Big\{ u \in H^2(\Omega, \mathbb{R}), \mathbf{Q} \in H^1(\Omega, S_0) \, ;$$

$$\mathbf{Q} = s\left(\mathbf{n} \otimes \mathbf{n} - \frac{\mathbf{I}_d}{d}\right) \text{ for some } s \in [0, 1], \mathbf{Q} = \mathbf{Q}_b \text{ on } \partial\Omega \Big\},$$

$$(8.14)$$

with $\mathbf{n} \in H^1(\Omega, \mathcal{S}^{d-1})$ and the Dirichlet boundary data $\mathbf{Q}_b \in H^{1/2}(\partial\Omega, S_0)$. For simplicity, we only consider Dirichlet boundary conditions for $\mathbf{Q}$ in this section, but other types of boundary conditions (e.g., a mixture of the Dirichlet and natural boundary conditions as used in Chapter 10) can be taken.

Notice that $f_n(\mathbf{Q}, \nabla\mathbf{Q})$ is the classical LdG model for nematic LC. It is a known result from Davis and Gartland (1998, Corollary 4.4) that there exists a minimiser of the functional $\int_\Omega f_n$ on $\mathbf{Q} \in H^1(\Omega, S_0)$ in three dimensions. Furthermore, Bedford (2014, Theorem 5.18) has given an existence result of the Ball Bodford model (8.8) for $\mathbf{Q} \in SBV(\Omega, S_0)$ and $u \in H^2(\Omega, \mathbb{R})$, also in three dimensions. Motivated by these two results, we can give the existence result of minimising our proposed free energy (8.10) via the direct method of calculus of variations (see e.g., Giaquinta, 1983, Section 3, Chapter 1) in the admissible space $\mathcal{A}^s$.

Theorem 8.2 (Existence of minimisers). *Let $\mathcal{J}$ be of the form (8.10) with positive parameters a_3, B, q, K, l. Then there exists a solution pair $(u^*, \mathbf{Q}^*)$ that minimises $\mathcal{J}$ over the admissible set $\mathcal{A}^s$.*

Proof. Note that both the smectic density f_s and the nematic bulk density f_n^b are bounded from below as $a_3, l > 0$. Thus, $\mathcal{J}$ is also bounded from below and we can choose a minimising sequence $\{(u_j, \mathbf{Q}_j)\}$, i.e.,

$$(u_j, \mathbf{Q}_j) \in \mathcal{A}^s, \quad \mathbf{Q}_j - \tilde{\mathbf{Q}} \in H_0^1(\Omega, S_0),$$
$$\mathcal{J}(u_j, \mathbf{Q}_j) \overset{j\to\infty}{\longrightarrow} \inf\{\mathcal{J}(u, \mathbf{Q}) : (u, \mathbf{Q}) \in \mathcal{A}^s, \ \mathbf{Q} - \tilde{\mathbf{Q}} \in H_0^1(\Omega, S_0)\} < \infty.$$
$$(8.15)$$

Here, we define $\tilde{\mathbf{Q}} \in H^1(\Omega, S_0)$ to be the extended function with trace $\mathbf{Q}_b$. We tackle the three terms in (8.10) separately in the following.

First, for the nematic energy term $\int_\Omega f_n(\mathbf{Q}, \nabla\mathbf{Q})$, we can follow the proof of Davis and Gartland (1998, Theorem 4.3) to obtain that $f_n(\mathbf{Q}_j, \nabla\mathbf{Q}_j)$ is coercive in $H^1(\Omega, S_0)$ in the sense that f_n grows unbounded as $\|\mathbf{Q}\|_1 \to \infty$, and thus the minimising sequence $\{\mathbf{Q}_j\}$ must be bounded. Since $H^1(\Omega)$ is a reflexive Banach Space, we have a subsequence (also denoted as $\{\mathbf{Q}_j\}$) that weakly converges to $\mathbf{Q}^* \in H^1(\Omega, S_0)$ such that $\mathbf{Q}^* - \tilde{\mathbf{Q}} \in H_0^1(\Omega)$, and from the Rellich–Kondrachov theorem it follows that:

$$\mathbf{Q}_j \to \mathbf{Q}^* \quad \text{in } L^2(\Omega),$$
$$\nabla\mathbf{Q}_j \rightharpoonup \nabla\mathbf{Q}^* \quad \text{in } L^2(\Omega).$$

The weakly lower semi-continuity of the nematic energy density f_n in (8.12) is guaranteed by Davis and Gartland (1998, Lemma 4.2), therefore,

$$\liminf_{j\to\infty} \int_\Omega f_n(\mathbf{Q}_j, \nabla\mathbf{Q}_j) \geq \int_\Omega f_n(\mathbf{Q}^*, \nabla\mathbf{Q}^*). \tag{8.16}$$

Next, for the smectic bulk term $\int_\Omega f_s(u)$, we can follow the proof in Bedford (2014, Theorem 5.19) with further details. By (8.15), we have

$$\sup_j \int_\Omega \left(\left|\mathcal{D}^2 u_j\right|^2 + |u_j|^2 \right) < \infty,$$

which implies an upper bound for ∇u_j using (Bedford, 2014, equation (5.42)):

$$\int_\Omega |\nabla v|^2 \leq C \left(\int_\Omega \left|\mathcal{D}^2 v\right|^2 + v^2 \right) \quad \forall v \in H^2(\Omega, \mathbb{R}).$$

Hence, $\{u_j\}$ is bounded in $H^2(\Omega)$ and thus there is a subsequence (also denoted as $\{u_j\}$) such that

$$u_j \rightharpoonup u^* \quad \text{in } H^2(\Omega).$$

Moreover, one can readily check that $\|u^*\|_\infty < \infty$ by the Sobolev embedding of $H^2(\Omega)$ into the Hölder spaces $\mathcal{C}^{\mathfrak{t},\varkappa_0}(\Omega)$ ($\mathfrak{t}+\varkappa_0 = 1$ for $d = 2$ and $\mathfrak{t}+\varkappa_0 = 1/2$ for $d = 3$) and the boundedness of domain Ω. Again, by the Rellich–Kondrachov theorem, we have

$$u_j \to u^* \quad \text{in } L^2(\Omega),$$
$$\mathcal{D}^2 u_j \rightharpoonup \mathcal{D}^2 u^* \quad \text{in } L^2(\Omega).$$

Noting that f_s is bounded from below for all $u \in H^2(\Omega)$, then there holds that

$$\liminf_{j\to\infty} \int_\Omega f_s(u_j) \geq \int_\Omega f_s(u^*). \tag{8.17}$$

Now, we consider the nematic-smectic coupling term in (8.10). Since the admissible space $\mathcal{A}^s$ admits uniaxial tensors, we calculate

$$|\mathbf{Q}_j|^2 = \left| s_j \left(\mathbf{n}_j \otimes \mathbf{n}_j - \frac{\mathbf{I}_d}{d} \right) \right|^2$$

$$= |s_j|^2 \left(|\mathbf{n}_j \otimes \mathbf{n}_j|^2 + \left|\frac{\mathbf{I}_d}{d}\right|^2 - \frac{2}{d}\mathbf{n}_j \otimes \mathbf{n}_j : \mathbf{I}_d \right)$$

$$= |s_j|^2 \left(1 + \frac{1}{d} - \frac{2}{d} \right)$$

$$= |s_j|^2 \left(1 - \frac{1}{d} \right)$$

$$< |s_j|^2,$$

implying that $|\mathbf{Q}_j|^2$ is always bounded in Ω. By this boundedness and the fact that $\|u^*\|_\infty < \infty$, we can deduce

$$\int_\Omega |u_j \mathbf{Q}_j - u^* \mathbf{Q}^*|^2 = \int_\Omega |(u_j - u^*)\mathbf{Q}_j + u^*(\mathbf{Q}_j - \mathbf{Q}^*)|^2$$

$$\leq 2 \int_\Omega \left(|u_j - u^*|^2 |\mathbf{Q}_j|^2 + |u^*|^2 |\mathbf{Q}_j - \mathbf{Q}^*|^2 \right)$$

$$\to 0 \quad \text{as } u_j \to u^*, \mathbf{Q}_j \to \mathbf{Q}^* \text{ in } L^2(\Omega).$$

Hence, $u_j \mathbf{Q}_j \to u^* \mathbf{Q}^*$ in $L^2(\Omega)$, and further,

$$u_j \left(\mathbf{Q}_j + \frac{\mathbf{I}_d}{d} \right) \to u^* \left(\mathbf{Q}^* + \frac{\mathbf{I}_d}{d} \right) \qquad \text{in } L^2(\Omega),$$

$$u_j \left(\mathbf{Q}_j + \frac{\mathbf{I}_d}{d} \right) : \mathcal{D}^2 u_j \rightharpoonup u^* \left(\mathbf{Q}^* + \frac{\mathbf{I}_d}{d} \right) : \mathcal{D}^2 u^* \quad \text{in } L^1(\Omega).$$

Therefore, we have

$$\liminf_{j \to \infty} \int_\Omega \left| \mathcal{D}^2 u_j + q^2 \left(\mathbf{Q}_j + \frac{\mathbf{I}_d}{d} \right) u_j \right|^2$$

$$= \liminf_{j \to \infty} \int_\Omega \left(|\mathcal{D}^2 u_j|^2 + 2q^2 u_j \left(\mathbf{Q}_j + \frac{\mathbf{I}_d}{d} \right) : \mathcal{D}^2 u_j + q^4 \left| u_j \left(\mathbf{Q}_j + \frac{\mathbf{I}_d}{d} \right) \right|^2 \right)$$

$$\geq \int_\Omega \left(|\mathcal{D}^2 u^*|^2 + 2q^2 u^* \left(\mathbf{Q}^* + \frac{\mathbf{I}_d}{d} \right) : \mathcal{D}^2 u^* + q^4 \left| u^* \left(\mathbf{Q}^* + \frac{\mathbf{I}_d}{d} \right) \right|^2 \right)$$

$$= \int_\Omega \left| \mathcal{D}^2 u^* + q^2 \left(\mathbf{Q}^* + \frac{\mathbf{I}_d}{d} \right) u^* \right|^2. \tag{8.18}$$

Finally, we only need to check that $\mathbf{Q}^*$ is uniaxial, i.e., $\mathbf{Q}^* = s^* \left(\mathbf{n}^* \otimes \mathbf{n}^* - \frac{\mathbf{I}_d}{d} \right)$ for certain s^* and $\mathbf{n}^*$. This is indeed guaranteed by the L^2 convergence of $\mathbf{Q}_j$ and the compactness of the unit sphere $\mathbf{n}_j$ lies. Hence, we can conclude that $\mathcal{J}(u^*, \mathbf{Q}^*)$ achieves its minimum in the admissible space $\mathcal{A}^s$ by combining (8.16), (8.17), and (8.18). $\qquad\square$

8.4 Summary

In this chapter, we reviewed three models for smectic-A LC: the classical dG model, a more recent model by Pevnyi, Selinger and Sluckin and the Ball–Bedford model. Through discussing their potential issues, it motivated us to propose a new model, incorporating the nematic tensor order parameter $\mathbf{Q}$ and a real smectic order parameter u, to characterise the complex defect structures existing in smectic liquid crystals. We then gave an existence result for the proposed model.

Chapter 9

Finite Element Discretisation

It is implied from Theorem 8.2 that there exist minimisers of the free energy functional (8.10). One might then naturally ask how those solutions behave. We therefore consider the discretisation of the problem in this chapter. For simplicity, we only consider the decoupled case, i.e., $q = 0$ where two separate problems are to be solved: a second-order PDE for the nematic tensor order parameter $\mathbf{Q}$ and a fourth-order PDE for the smectic density variation u. With the derived *a priori* error estimates at hand, we then choose a suitable finite element pair for $(\mathbf{Q}, u)$, to be used in the implementations of some realistic scenarios as illustrated in Chapter 10. We verify the expected convergence behaviour via the manufactured method of solutions for both $q = 0$ and $q > 0$.

9.1 Some Basics About Finite Elements

To solve the derived minimisation problem above, the *finite element method* (FEM) is the most conventional and commonly used technique, which relies on an essential ingredient, i.e., the weak formulation of the problem. One may refer to Braess and Schumaker (1997) for more information on FEM especially in the realm of solid mechanics.

Now, the finite element solution to the variational problem should be sought. To do so, we start with a standard finite element discretisation by partitioning the domain into nonoverlapping elements and defining a set of local basis functions on each element, where we have used the continuous linear nodal basis functions in our implementation. Then by the requirement of continuity across all the faces in the global setting, the piecewise linear global basis functions are presented. Next, we can approximate $H^1(\mathcal{B}_0; \mathbb{R}^3)$ by the finite element space spanned from all these global basis functions with the unknowns to be their coefficients, which are also known as degrees of freedom. Here, we ignore further details for constructing finite element approximation but one could easily find some fully described references on FEM (to name a few, Brenner and Scott, 2008; Ciarlet, 2002).

At this stage, the finite element solution can be represented as a linear combination of the global basis functions and after substituting it into the variational form, we are able to obtain an algebraic system (generally nonlinear) denoted by

$$\mathbf{A}(\mathbf{u}) = \mathbf{f}, \tag{9.1}$$

where $\mathbf{u}$ is the solution.

9.2   *A Priori* Analysis for $q = 0$

In the decoupled case, we are to solve two independent minimisation problems: one for the tensor field $\mathbf{Q}$,

$$\min_{\mathbf{Q} \in H_b^1(\Omega, S_0)} \quad \mathcal{J}_1(\mathbf{Q}) = \int_\Omega \left(f_n(\mathbf{Q}, \nabla \mathbf{Q}) \right),$$

and the other for the density variation u:

$$\min_{u \in H^2(\Omega, \mathbb{R})} \quad \mathcal{J}_2(u) = \int_\Omega \left(B \left| \mathcal{D}^2 u \right|^2 + f_s(u) \right).$$

One can derive the following strong forms of their equilibrium equations using integration by parts (and assuming that $u \in H^4(\Omega)$),

$$(\mathcal{P}1) \quad \begin{cases} d = 2 \Rightarrow -K \Delta \mathbf{Q} + 2l \left(2|\mathbf{Q}|^2 - 1 \right) \mathbf{Q} = 0 & \text{in } \Omega, \\ d = 3 \Rightarrow -K \Delta \mathbf{Q} + l \left(-\mathbf{Q} - |\mathbf{Q}|^2 + 2|\mathbf{Q}|^2 \mathbf{Q} \right) = 0 & \text{in } \Omega, \\ \mathbf{Q} = \mathbf{Q}_b & \text{on } \partial\Omega, \end{cases} \tag{9.2}$$

and

$$\begin{cases} 2B \left(\mathcal{D}^2 : \mathcal{D}^2 \right) u + a_1 u + a_2 u^2 + a_3 u^3 = 0 & \text{in } \Omega, \\ S_{bc}^0(u; v) = 0 \ \forall v \in H^2(\Omega) & \text{on } \partial\Omega, \end{cases} \tag{9.3}$$

with the natural boundary data given by

$$S_{bc}^0(u; v) := \int_{\partial\Omega} \left\{ \nu \cdot \left(\mathcal{D}^2 u \cdot \nabla v \right) - \left((\nabla \cdot \mathcal{D}^2 u) \cdot \nu \right) v \right\}.$$

Note that we are not enforcing any essential boundary conditions for the real variable u in (9.3). This can be insufficient to guarantee the uniqueness of solutions, thus leading to ill-posed problems. In fact, both u and $-u$ are admissible solutions if $a_2 = 0$. Moreover, one expect the solution u of the smectic-A model (8.10) to be a cosine function that describes the periodicity as illustrated in Pevnyi *et al.* (2014, Eq. (5)) due to the alignment between smectic layer normals and directors. Therefore, the lack of essential boundary conditions may result in multiple solutions with shifted phases.

To facilitate our analysis, we assume that the fourth order problem is imposed with a Dirichlet boundary condition $u = u_b$ on $\partial\Omega$ and a natural boundary condition regarding the second derivative of u. That is to say, we consider the following minimisation problem for u:

$$\min_{u \in H^2 \cap H_b^1(\Omega, \mathbb{R})} \quad \mathcal{J}_2(u) = \int_\Omega \left(B \left| \mathcal{D}^2 u \right|^2 + f_s(u) \right),$$

which corresponds to a strong form

$$(\mathcal{P}2) \quad \begin{cases} 2B \left(\mathcal{D}^2 : \mathcal{D}^2 \right) u + a_1 u + a_2 u^2 + a_3 u^3 = 0 & \text{in } \Omega, \\ u = u_b & \text{on } \partial\Omega, \\ \mathcal{D}^2 u \cdot \nu = \mathcal{D}^2 u_b \cdot \nu & \text{on } \partial\Omega, \end{cases} \quad (9.4)$$

where $H_b^1(\Omega, \mathbb{R}) := \{v \in H^1(\Omega, \mathbb{R}) : v = u_b \text{ on } \partial\Omega\}$.

Remark 9.1. The uniqueness result of the problem (9.4) is still not guaranteed, though we have imposed additional boundary conditions. This can be resulted from the presence of the nonlinear term.

Remark 9.2. In the coupled case that we implement in Chapter 10, the boundary conditions on u are somewhat different. No essential boundary conditions are enforced, and some second derivative terms arise in the natural boundary condition. See (A.5) for details.

Essentially, $(\mathcal{P}1)$ is a second-order semi-linear PDE while $(\mathcal{P}2)$ yields a fourth order semi-linear PDE. To be more specific, both PDEs possess cubic nonlinearities. We now consider these two problems separately.

9.2.1 *A priori error estimates for* $(\mathcal{P}1)$

Note that problem $(\mathcal{P}1)$ is a special form of the classical Landau–de Gennes model of nematic liquid crystals. Finite element analysis for a more general form using conforming discretisations has been studied in Davis and Gartland (1998) with homogeneous Dirichlet boundary data and in Davis (1994) with inhomogeneous Dirichlet and natural boundary conditions. More specifically, Davis and Gartland (1998) gave an abstract non-linear finite element convergence analysis where an optimal H^1 error bound is proved on convex domains with piecewise linear polynomial approximations. However, the L^2 error bound is not derived. Quite recently, Maity *et al.* (2020) analysed the discontinuous Galerkin finite element methods (dGFEM) for a two-dimensional reduced Landau–de Gennes free energy, where optimal *a priori* error estimates in the L^2-norm with exact solutions being in H^2 and piecewise linear polynomial approximations are achieved. Their representations of the nonlinear variational form and approaches of deriving error estimates are different from those of Davis and Gartland. We follow similar techniques from Maity *et al.* (2020) in this subsection for concreteness.

We use the common continuous Lagrange elements for the problem ($\mathcal{P}1$). For simplicity, we only illustrate the analysis in two dimensions for the model problem (8.10); the three-dimensional case has an additional quadratic term $|\mathbf{Q}|^2$ in the strong form which can be tackled similarly. Since ($\mathcal{P}1$) arises in the classical LdG model for nematic LC, we can quote some existing results (e.g., regularity, convergence rate in the H^1 norm).

Theorem 9.1 (Davis and Gartland, 1998, Theorem 6.3; Regularity).   *Let Ω be an open, bounded, Lipschitz and convex domain. If the Dirichlet data $\mathbf{Q}_b \in H^{1/2}(\partial\Omega, S_0)$, then any solution of ($\mathcal{P}1$) belongs to $H^2(\Omega, S_0)$.*

Remark 9.3. One may wonder that the H^2-regularity of $\mathbf{Q}$ possibly excludes the appearance of singularities, e.g., the half-charge defects in nematics. Indeed, we do not consider the case with singularities in the analysis throughout this part of work.

Suppose $\mathbf{Q}_h \in \mathbf{V}_h$ is the approximate solution (of the discrete problem (9.6) introduced later) by finite element methods on a finite-dimensional space $\mathbf{V}_h \subset H_b^1(\Omega, S_0)$. For simplicity, we restrict ourselves in the case that $\mathbf{V}_h$ consists of piecewise linear polynomials. An *a priori* estimate in the H^1 norm has been shown in Davis (1994, Theorem 2.3.3) and Davis and Gartland (1998, Theorem 7.3) and we include it here for self-containment.

Theorem 9.2 (Davis and Gartland, 1998, Theorem 7.3; H^1 error estimate for Q). *Let Ω be an open, bounded, polygonal and convex domain. If $\mathbf{Q} \in H^2 \cap H_b^1(\Omega, S_0)$ and $\mathbf{Q}_h \in \mathbf{V}_h$ represents an approximated solution to $\mathbf{Q}$, it holds that*

$$\|\mathbf{Q} - \mathbf{Q}_h\|_1 \lesssim h\|\mathbf{Q}\|_2. \tag{9.5}$$

Remark 9.4. Theorems 9.1 and 9.2 hold for both $\Omega \subset \mathbb{R}^2$ and $\Omega \subset \mathbb{R}^3$.

Following the same representation of the nonlinear variational form as in [Maity *et al.* (2020)], we introduce the continuous weak formulation of $(\mathcal{P}1)$: find $\mathbf{Q} \in H_b^1(\Omega, S_0)$ such that

$$\mathcal{N}^n(\mathbf{Q})\mathbf{P} := A^n(\mathbf{Q}, \mathbf{P}) + B^n(\mathbf{Q}, \mathbf{Q}, \mathbf{Q}, \mathbf{P}) + C^n(\mathbf{Q}, \mathbf{P}) = 0 \quad \forall \mathbf{P} \in \mathbf{H}_0^1(\Omega), \tag{9.6}$$

where the bilinear forms are

$$A^n(\mathbf{Q}, \mathbf{P}) := K \int_\Omega \nabla\mathbf{Q} : \nabla\mathbf{P},$$

$$C^n(\mathbf{Q}, \mathbf{P}) := -2l \int_\Omega \mathbf{Q} : \mathbf{P},$$

and the nonlinear operator is given by

$$B^n(\Psi, \Phi, \Theta, \Xi) := \frac{4l}{3} \int_\Omega \left((\Psi : \Phi)(\Theta : \Xi) + 2(\Psi : \Theta)(\Phi : \Xi) \right). \tag{9.7}$$

Since (9.6) is nonlinear, we need to approximate the solution of its linearised version, i.e., find $\Theta \in \mathbf{H}_0^1(\Omega)$ such that

$$\langle \mathcal{D}\mathcal{N}^n(\mathbf{Q})\Theta, \Phi \rangle := A^n(\Theta, \Phi) + 3B^n(\mathbf{Q}, \mathbf{Q}, \Theta, \Phi) + C^n(\Theta, \Phi)$$
$$= -\mathcal{N}^n(\mathbf{Q})\Phi \quad \forall \Phi \in \mathbf{H}_0^1(\Omega), \tag{9.8}$$

where $\langle \cdot, \cdot \rangle$ represents the dual pairing between $\mathbf{H}^{-1}(\Omega)$ and $\mathbf{H}_0^1(\Omega)$. We use continuous Lagrange elements and the finite-dimensional approximation space $\mathbf{V}_h \subset \mathbf{H}^1(\Omega)$, thus, the discrete bilinear form inherits from (9.8).

Remark 9.5. We only consider the approximation of a regular or non-singular solution $\mathbf{Q}$ of (9.6). This means that the Implicit Function Theorem can be applied in the Banach space $\mathbf{H}^1(\Omega)$ and it is equivalent to the following continuous inf-sup condition (Maity *et al.*, 2020, Equation (2.8)):

$$0 < \beta_Q := \inf_{\substack{\Theta \in \mathbf{H}^1(\Omega) \\ \|\Theta\|_1 = 1}} \sup_{\substack{\Phi \in \mathbf{H}^1(\Omega) \\ \|\Phi\|_1 = 1}} \langle \mathcal{D}\mathcal{N}^n(\mathbf{Q})\Theta, \Phi \rangle$$

$$= \inf_{\substack{\Phi \in \mathbf{H}^1(\Omega) \\ \|\Phi\|_1 = 1}} \sup_{\substack{\Theta \in \mathbf{H}^1(\Omega) \\ \|\Theta\|_1 = 1}} \langle \mathcal{D}\mathcal{N}^n(\mathbf{Q})\Theta, \Phi \rangle. \tag{9.9}$$

To deduce the L^2 error estimate of regular solutions one can use the Aubin–Nitsche duality argument; however due to the nonlinearity, it is nontrivial to derive the dual problem. To this end, we consider the following linear dual problem to the primary nonlinear problem (9.2): find $\mathbf{N} \in \mathbf{H}_0^1(\Omega)$ such that

$$\begin{cases} -K\Delta\mathbf{N} + 4l|\mathbf{Q}|^2\mathbf{N} + 8l(\mathbf{Q}:\mathbf{N})\mathbf{Q} - 2l\mathbf{N} = \mathbf{G} & \text{in } \Omega, \\ \mathbf{N} = \mathbf{0} & \text{on } \partial\Omega, \end{cases} \tag{9.10}$$

for a given $\mathbf{G} \in \mathbf{L}^2(\Omega)$ (we will see the choice of $\mathbf{G}$ in the proof of Theorem 9.8). Here, $\mathbf{Q} \in \mathbf{H}_b^1(\Omega)$. Furthermore, one can obtain the weak form of (9.10): find $\mathbf{N} \in \mathbf{H}_0^1(\Omega)$ such that

$$\langle \mathcal{D}\mathcal{N}^n(\mathbf{Q})\mathbf{N}, \Phi \rangle = A^n(\mathbf{N}, \Phi) + 3B^n(\mathbf{Q}, \mathbf{Q}, \mathbf{N}, \Phi) + C^n(\mathbf{N}, \Phi) = (\mathbf{G}, \Phi)_0. \tag{9.11}$$

The technique follows (Maity *et al.*, 2020), where their proofs based on dGFEM with the broken Sobolev space

$$\mathbf{H}_0^1(\mathcal{T}_h) = \left\{ \mathbf{v} \in \mathbf{L}^2(\Omega) : \mathbf{v}|_T \in \mathbf{H}^1(T) \ \forall T \in \mathcal{T}_h, \mathbf{v} = \mathbf{0} \text{ on } \partial\Omega \right\},$$

are derived with the mesh-dependent norm

$$\|\mathbf{v}\|_{dG}^2 = \sum_{T \in \mathcal{T}_h} \int_T |\nabla\mathbf{v}|^2 + \sum_{e \in \mathcal{E}} \int_e \frac{\sigma_m}{h_e} [\![\mathbf{v}]\!]^2.$$

Here, $\sigma_m > 0$ is the penalty parameter. Moreover, for any interior edge $e \in \mathcal{E}_I$ shared by two cells T_- and T_+, we define the jump $[\![\mathbf{v}]\!]$ by $[\![\mathbf{v}]\!] = \mathbf{v}_- \cdot \nu_- + \mathbf{v}_+ \cdot \nu_+$ with ν_-, ν_+ representing the restriction of outward normals in T_-, T_+, respectively. On the boundary edge/face $e \in \mathcal{E}_B$, we define $[\![\mathbf{v}]\!] = \mathbf{v} \cdot \nu$.

One can easily check that for a continuous approximation $\mathbf{v}_h \in \mathbf{H}_0^1(\Omega)$, it holds that $[\![\mathbf{v}_h]\!] = 0$ and the $\|\cdot\|_{dG}$-norm is in fact the $\mathbf{H}^1$ semi-norm in the Sobolev space $\mathbf{H}^1(\Omega)$ and equivalent to the $\|\cdot\|_1$-norm in the Sobolev space $\mathbf{H}_0^1(\Omega)$ by the Poincaré inequality. Hence, it is straightforward to derive similar results for the $\|\cdot\|_1$-norm as in Maity *et al.* (2020). We give some auxiliary results about the operators $A^n(\cdot, \cdot)$, $B^n(\cdot, \cdot, \cdot, \cdot)$ and $C^n(\cdot, \cdot)$.

Lemma 9.3 (Boundedness and coercivity of $A^n(\cdot, \cdot)$). *For* $\Theta, \Phi \in \mathbf{H}_0^1(\Omega)$, *there holds*

$$A^n(\Theta, \Phi) \lesssim \|\Theta\|_1 \|\Phi\|_1,$$

and

$$\|\Theta\|_1^2 \lesssim A^n(\Theta, \Theta) \quad \forall \Theta \in \mathbf{H}_0^1(\Omega).$$

Proof. An application of the Cauchy–Schwarz inequality yields the boundedness result while the coercivity follows from the Poincaré inequality. □

Lemma 9.4 (Boundedness of $B^n(\cdot, \cdot, \cdot, \cdot)$, $C^n(\cdot, \cdot)$). *For* $\Psi, \Phi, \Theta, \Xi \in \mathbf{H}^1(\Omega)$, *there holds*

$$B^n(\Psi, \Phi, \Theta, \Xi) \lesssim \|\Psi\|_1 \|\Phi\|_1 \|\Theta\|_1 \|\Xi\|_1, \quad C^n(\Psi, \Phi) \lesssim \|\Psi\|_1 \|\Phi\|_1, \quad (9.12)$$

and for $\Psi, \Phi \in \mathbf{H}^2(\Omega)$, $\Theta, \Xi \in \mathbf{H}^1(\Omega)$,

$$B^n(\Psi, \Phi, \Theta, \Xi) \lesssim \|\Psi\|_2 \|\Phi\|_2 \|\Theta\|_1 \|\Xi\|_1. \quad (9.13)$$

Proof. For $\Psi, \Phi, \Theta, \Xi \in \mathbf{H}^1(\Omega)$, we use Hölder's inequality and the embedding result $\mathbf{H}^1(\Omega) \hookrightarrow \mathbf{L}^4(\Omega)$ to obtain

$$B^n(\Psi, \Phi, \Theta, \Xi) \lesssim \|\Psi\|_{L^4} \|\Phi\|_{L^4} \|\Theta\|_{L^4} \|\Xi\|_{L^4} \lesssim \|\Psi\|_1 \|\Phi\|_1 \|\Theta\|_1 \|\Xi\|_1.$$

The proof of (9.13) follows analogously to that of (9.12) with the use of the embedding result $\mathbf{H}^2(\Omega) \hookrightarrow \mathbf{L}^\infty(\Omega)$ and Cauchy–Schwarz inequality:

$$B^n(\Psi, \Phi, \Theta, \Xi) \lesssim \|\Psi\|_\infty \|\Phi\|_\infty \|\Theta\|_0 \|\Xi\|_0 \lesssim \|\Psi\|_2 \|\Phi\|_2 \|\Theta\|_1 \|\Xi\|_1.$$

This completes the proof. □

We also quote interpolation estimates that will be frequently used.

Lemma 9.5 (Brenner and Scott, 2008; Interpolation estimates). *For* $\mathbf{v} \in \mathbf{H}^2(\Omega)$ *there exists* $I_h \mathbf{v} \in \mathbf{V}_h$ *such that*

$$\|\mathbf{v} - I_h \mathbf{v}\|_0 \lesssim h^2 \|\mathbf{v}\|_2,$$

$$\|\mathbf{v} - I_h \mathbf{v}\|_1 \lesssim h \|\mathbf{v}\|_2.$$

Here, $I_h : \mathbf{H}^2 \to \mathbf{V}_h$ *is the interpolation operator.*

To derive the L^2 *a priori* error estimates, we need two more auxiliary results.

Lemma 9.6. *For* $\mathbf{Q} \in \mathbf{H}^2(\Omega) \cap \mathbf{H}^1_b(\Omega)$, $\mathbf{N} \in \mathbf{H}^2(\Omega) \cap \mathbf{H}^1_0(\Omega)$ *and* $I_h\mathbf{Q} \in \mathbf{V}_h \subset \mathbf{H}^1_b(\Omega)$, *it holds that*

$$A^n(I_h\mathbf{Q} - \mathbf{Q}, \mathbf{N}) \lesssim h^2 \|\mathbf{Q}\|_2 \|\mathbf{N}\|_2.$$

Proof. By the definition of the bilinear form $A^n(\cdot, \cdot)$, integration by parts (note that $(I_h\mathbf{Q}-\mathbf{Q})|_{\partial\Omega} = 0$), Cauchy–Schwarz inequality and interpolation estimates from Lemma 9.5, we have

$$A^n(I_h\mathbf{Q} - \mathbf{Q}, \mathbf{N}) = \int_\Omega K\nabla(I_h\mathbf{Q} - \mathbf{Q}) \cdot \nabla\mathbf{N}$$

$$= -\int_\Omega K(I_h\mathbf{Q} - \mathbf{Q}) \cdot \Delta\mathbf{N}$$

$$\lesssim \|I_h\mathbf{Q} - \mathbf{Q}\|_0 \|\mathbf{N}\|_2$$

$$\lesssim h^2 \|\mathbf{Q}\|_2 \|\mathbf{N}\|_2.$$

This completes the proof. □

We then show that the H^2-norm of the dual solution is bounded by the source term $\mathbf{G} \in \mathbf{L}^2(\Omega)$.

Lemma 9.7 (Boundedness of the dual solution in the H^2-norm). *The solution* $\mathbf{N}$ *to the weak form* (9.11) *of the dual linear problem belongs to* $\mathbf{H}^2(\Omega) \cap \mathbf{H}^1_0(\Omega)$ *and it holds that*

$$\|\mathbf{N}\|_2 \lesssim \|\mathbf{G}\|_0. \tag{9.14}$$

Proof. We use the inf-sup condition (9.9) for the linear operator $\langle \mathcal{D}\mathcal{N}^n(\mathbf{Q})\cdot, \cdot \rangle$, the weak formulation (9.11) and Cauchy–Schwarz inequality to obtain

$$\beta_Q \|\mathbf{N}\|_1 \leq \sup_{\substack{\Phi \in \mathbf{H}^1_0 \\ \|\Phi\|_1 = 1}} \langle \mathcal{D}\mathcal{N}^n(\mathbf{Q})\mathbf{N}, \Phi \rangle = \sup_{\substack{\Phi \in \mathbf{H}^1_0 \\ \|\Phi\|_1 = 1}} (\mathbf{G}, \Phi)_0 \leq \|\mathbf{G}\|_0. \tag{9.15}$$

By the form of (9.11) and boundedness of $B^n(\mathbf{Q}, \mathbf{Q}, \cdot, \cdot)$ and $C^n(\cdot, \cdot)$, we have

$$\|K\Delta\mathbf{N}\|_0 = \| -3B^n(\mathbf{Q}, \mathbf{Q}, \cdot, \mathbf{N}) - C^n(\cdot, \mathbf{N}) + (\mathbf{G}, \cdot)_0\|_0 \lesssim \|\mathbf{N}\|_1 + \|\mathbf{G}\|_0. \tag{9.16}$$

Note that the linear dual problem (9.10) includes a Laplace operator. Using the elliptic regularity result on a domain with polygonal boundary (see, e.g., Grisvard, 1985, Theorem 4.3.1.4) for Laplace operators, we deduce that $\mathbf{N} \in \mathbf{H}^2(\Omega)$. Combining Equations (9.15) and (9.16) and the fact that $\|\Delta \cdot \|_0$ is indeed a norm in $\mathbf{H}^2(\Omega) \cap \mathbf{H}_0^1(\Omega)$, we can get (9.14). □

Finally, we are ready to deduce the optimal L^2 error estimate.

Theorem 9.8 (L^2 error estimate). *Let $\mathbf{Q}$ be a regular solution of the nonlinear weak problem (9.6). For sufficiently small mesh size h, there exists a unique approximate solution $\mathbf{Q}_h$ of the discrete problem (having the same weak formulation as (9.6)) such that*

$$\|\mathbf{Q} - \mathbf{Q}_h\|_0 \lesssim h^2 \left(2 + \left(3 + 2h + 2h^2\right) \|\mathbf{Q}\|_2^2\right) \|\mathbf{Q}\|_2. \tag{9.17}$$

Proof. We take $\mathbf{G} = I_h\mathbf{Q} - \mathbf{Q}_h$ in the linear dual problem (9.10), multiply (9.10) by $I_h\mathbf{Q} - \mathbf{Q}_h$ and integrate by parts to obtain the weak formulation

$$\langle \mathcal{D}\mathcal{N}^n(\mathbf{Q})(I_h\mathbf{Q} - \mathbf{Q}_h), \mathbf{N} \rangle = \|I_h\mathbf{Q} - \mathbf{Q}_h\|_0^2.$$

Here, $\langle \mathcal{D}\mathcal{N}^n(\mathbf{Q})(I_h\mathbf{Q} - \mathbf{Q}_h), \mathbf{N} \rangle = A^n(I_h\mathbf{Q} - \mathbf{Q}_h, \mathbf{N}) + 3B^n(\mathbf{Q}, \mathbf{Q}, I_h\mathbf{Q} - \mathbf{Q}_h, \mathbf{N}) + C^n(I_h\mathbf{Q} - \mathbf{Q}_h, \mathbf{N})$. Since both $\mathbf{Q}$ and its approximation $\mathbf{Q}_h$ satisfy the weak formulation (9.6), we know

$$\mathcal{N}^n(\mathbf{Q})I_h\mathbf{N} = 0 \quad \text{and} \quad \mathcal{N}^n(\mathbf{Q}_h)I_h\mathbf{N} = 0.$$

By the definitions of the nonlinear operator $\mathcal{N}^n(\mathbf{Q})\cdot$ and bilinear form $\langle \mathcal{D}\mathcal{N}^n(\mathbf{Q})\cdot, \cdot \rangle$, we calculate

$$\|I_h\mathbf{Q} - \mathbf{Q}_h\|_0^2 \tag{9.18}$$

$$= \langle \mathcal{D}\mathcal{N}^n(\mathbf{Q})(I_h\mathbf{Q} - \mathbf{Q}_h), \mathbf{N} \rangle + \mathcal{N}^n(\mathbf{Q}_h)I_h\mathbf{N} - \mathcal{N}^n(\mathbf{Q})I_h\mathbf{N}$$

$$= A^n(I_h\mathbf{Q} - \mathbf{Q}_h, \mathbf{N}) + 3B^n(\mathbf{Q}, \mathbf{Q}, I_h\mathbf{Q} - \mathbf{Q}_h, \mathbf{N}) + C^n(I_h\mathbf{Q} - \mathbf{Q}_h, \mathbf{N})$$

$$\quad + A^n(\mathbf{Q}_h, I_h\mathbf{N}) + B^n(\mathbf{Q}_h, \mathbf{Q}_h, \mathbf{Q}_h, I_h\mathbf{N}) + C^n(\mathbf{Q}_h, I_h\mathbf{N})$$

$$\quad - A^n(\mathbf{Q}, I_h\mathbf{N}) - B^n(\mathbf{Q}, \mathbf{Q}, \mathbf{Q}, I_h\mathbf{N}) - C^n(\mathbf{Q}, I_h\mathbf{N})$$

$$= \underbrace{A^n(I_h\mathbf{Q} - \mathbf{Q}, \mathbf{N}) + A^n(\mathbf{Q} - \mathbf{Q}_h, \mathbf{N} - I_h\mathbf{N})}_{U_1}$$

$$\quad + \underbrace{C^n(I_h\mathbf{Q} - \mathbf{Q}, \mathbf{N}) + C^n(\mathbf{Q} - \mathbf{Q}_h, \mathbf{N} - I_h\mathbf{N})}_{U_2}$$

$$+ \underbrace{3B^n(\mathbf{Q}, \mathbf{Q}, I_h\mathbf{Q} - \mathbf{Q}_h, \mathbf{N} - I_h\mathbf{N}) + 3B^n(\mathbf{Q}, \mathbf{Q}, I_h\mathbf{Q} - \mathbf{Q}, I_h\mathbf{N})}_{U_3}$$

$$+ \underbrace{B^n(\mathbf{Q}_h, \mathbf{Q}_h, \mathbf{Q}_h, I_h\mathbf{N}) - 3B^n(\mathbf{Q}, \mathbf{Q}, \mathbf{Q}_h, I_h\mathbf{N}) + 2B^n(\mathbf{Q}, \mathbf{Q}, \mathbf{Q}, I_h\mathbf{N})}_{U_4}$$

$$=: U_1 + U_2 + U_3 + U_4. \tag{9.19}$$

We now use the previous auxiliary results to bound $U_1, \ldots, U_4$ separately, yielding

$$
\begin{aligned}
U_1 &= A^n(I_h\mathbf{Q} - \mathbf{Q}, \mathbf{N}) + A^n(\mathbf{Q} - \mathbf{Q}_h, \mathbf{N} - I_h\mathbf{N}) \\
&\lesssim h^2\|\mathbf{Q}\|_2\|\mathbf{N}\|_2 + \|\mathbf{Q} - \mathbf{Q}_h\|_1\|\mathbf{N} - I_h\mathbf{N}\|_1, \quad \text{by Lemmas 9.3 and 9.6,} \\
&\lesssim h^2\|\mathbf{Q}\|_2\|\mathbf{N}\|_2, \quad\quad\quad\quad\quad\quad\quad\quad\quad \text{by (9.5) and Lemma 9.5,}
\end{aligned}
\tag{9.20}
$$

$$
\begin{aligned}
U_2 &= C^n(I_h\mathbf{Q} - \mathbf{Q}, \mathbf{N}) + C^n(\mathbf{Q} - \mathbf{Q}_h, \mathbf{N} - I_h\mathbf{N}) \\
&\lesssim \|I_h\mathbf{Q} - \mathbf{Q}\|_0\|\mathbf{N}\|_0 + \|\mathbf{Q} - \mathbf{Q}_h\|_1\|\mathbf{N} - I_h\mathbf{N}\|_1, \quad \text{by CS and (9.12),} \\
&\lesssim h^2\|\mathbf{Q}\|_2(\|\mathbf{N}\|_0 + \|\mathbf{N}\|_2), \quad\quad\quad\quad\quad \text{by Lemma 9.5 and (9.5),} \\
&\lesssim h^2\|\mathbf{Q}\|_2\|\mathbf{N}\|_2, \quad\quad\quad\quad\quad\quad\quad\quad\quad \text{by } \|\mathbf{N}\|_0 \leq \|\mathbf{N}\|_2,
\end{aligned}
\tag{9.21}
$$

and

$$
\begin{aligned}
U_3 &= 3B^n(\mathbf{Q}, \mathbf{Q}, I_h\mathbf{Q} - \mathbf{Q}_h, \mathbf{N} - I_h\mathbf{N}) + 3B^n(\mathbf{Q}, \mathbf{Q}, I_h\mathbf{Q} - \mathbf{Q}, I_h\mathbf{N}) \\
&\lesssim \|\mathbf{Q}\|_2^2\|I_h\mathbf{Q} - \mathbf{Q}_h\|_1\|\mathbf{N} - I_h\mathbf{N}\|_1 + \|\mathbf{Q}\|_2^2\|I_h\mathbf{Q} - \mathbf{Q}\|_0\|I_h\mathbf{N}\|_0, \\
&\quad \text{by (9.13) and CS,} \\
&\lesssim h\|\mathbf{Q}\|_2^2\|I_h\mathbf{Q} - \mathbf{Q}_h\|_1\|\mathbf{N}\|_2 + h^2\|\mathbf{Q}\|_2^3\|I_h\mathbf{N}\|_0, \\
&\quad \text{by Lemma 9.5.}
\end{aligned}
\tag{9.22}
$$

Here, CS abbreviates for the Cauchy–Schwarz inequality. Note that by triangle inequality, Lemma 9.5 and (9.5), it holds that

$$
\begin{aligned}
\|I_h\mathbf{Q} - \mathbf{Q}_h\|_1 &\leq \|I_h\mathbf{Q} - \mathbf{Q}\|_1 + \|\mathbf{Q} - \mathbf{Q}_h\|_1 \lesssim h\|\mathbf{Q}\|_2, \\
\|I_h\mathbf{N}\|_0 &\leq \|I_h\mathbf{N} - \mathbf{N}\|_0 + \|\mathbf{N}\|_0 \lesssim (1 + h^2)\|\mathbf{N}\|_2, \\
\|I_h\mathbf{N}\|_1 &\leq \|I_h\mathbf{N} - \mathbf{N}\|_1 + \|\mathbf{N}\|_1 \lesssim (1 + h)\|\mathbf{N}\|_2.
\end{aligned}
\tag{9.23}
$$

Therefore, we further estimate U_3 in (9.22) to obtain

$$U_3 \lesssim h^2(2 + h^2)\|\mathbf{Q}\|_2^3\|\mathbf{N}\|_2. \tag{9.24}$$

It remains to bound the U_4 term in (9.19). Let $\mathbf{E} = \mathbf{Q}_h - \mathbf{Q}$. We use the definition (9.7) of $B^n(\cdot, \cdot)$ and manipulate terms as follows:

$$U_4 = B^n(\mathbf{Q}_h, \mathbf{Q}_h, \mathbf{Q}_h, I_h\mathbf{N}) - 3B^n(\mathbf{Q}, \mathbf{Q}, \mathbf{Q}_h, I_h\mathbf{N}) + 2B^n(\mathbf{Q}, \mathbf{Q}, \mathbf{Q}, I_h\mathbf{N})$$

$$= 4l \int_\Omega |\mathbf{Q}_h|^2(\mathbf{Q}_h : I_h\mathbf{N}) - 4l \int_\Omega (|\mathbf{Q}|^2(\mathbf{Q}_h : I_h\mathbf{N}) + 2(\mathbf{Q} : \mathbf{Q}_h)(\mathbf{Q} : I_h\mathbf{N}))$$

$$+ 8l \int_\Omega |\mathbf{Q}|^2(\mathbf{Q} : I_h\mathbf{N})$$

$$= 4l \int_\Omega (|\mathbf{Q}_h|^2 - |\mathbf{Q}|^2)(\mathbf{Q}_h : I_h\mathbf{N}) - 8l \int_\Omega \mathbf{Q} : (\mathbf{Q}_h - \mathbf{Q})(\mathbf{Q} : I_h\mathbf{N})$$

$$= 4l \int_\Omega \mathbf{E} : (\mathbf{E} + 2\mathbf{Q})(\mathbf{E} + \mathbf{Q}) : I_h\mathbf{N} - 8l \int_\Omega (\mathbf{E} : \mathbf{Q})(\mathbf{Q} : I_h\mathbf{N})$$

$$= 4l \int_\Omega (\mathbf{E} : \mathbf{E})(\mathbf{E} : I_h\mathbf{N}) + 4l \int_\Omega (\mathbf{E} : \mathbf{E})(\mathbf{Q} : I_h\mathbf{N}) + 8l \int_\Omega (\mathbf{E} : \mathbf{Q})(\mathbf{E} : I_h\mathbf{N}).$$

Using the Hölder's inequality and the embedding result $\mathbf{H}^1(\Omega) \hookrightarrow \mathbf{L}^4(\Omega)$, we can bound U_4 further to obtain

$$
\begin{aligned}
U_4 &\lesssim \|\mathbf{E}\|_1^3\|I_h\mathbf{N}\|_1 + \|\mathbf{E}\|_1^2\|\mathbf{Q}\|_1\|I_h\mathbf{N}\|_1 + \|\mathbf{E}\|_1^2\|\mathbf{Q}\|_1\|I_h\mathbf{N}\|_1 \\
&= \|\mathbf{E}\|_1^2(\|\mathbf{E}\|_1 + 2\|\mathbf{Q}\|_1)\|I_h\mathbf{N}\|_1 \\
&\lesssim (1 + h)\|\mathbf{E}\|_1^2(\|\mathbf{E}\|_1 + \|\mathbf{Q}\|_1)\|\mathbf{N}\|_2, \qquad\qquad \text{by (9.23)}, \\
&\lesssim h^2(1 + h)\|\mathbf{Q}\|_2^2(h\|\mathbf{Q}\|_2 + \|\mathbf{Q}\|_2)\|\mathbf{N}\|_2, \qquad \text{by Lemma 9.5}, \\
&= h^2(1 + h)^2\|\mathbf{Q}\|_2^3\|\mathbf{N}\|_2. \tag{9.25}
\end{aligned}
$$

Combining the estimates (9.20), (9.21), (9.24), and (9.25) and applying Lemma 9.7 yields

$$
\begin{aligned}
\|I_h\mathbf{Q} - \mathbf{Q}_h\|_0^2 &\lesssim h^2\left(2 + \left(2 + h^2 + (1 + h)^2\right)\|\mathbf{Q}\|_2^2\right)\|\mathbf{Q}\|_2\|\mathbf{N}\|_2 \\
&\lesssim h^2\left(2 + \left(3 + 2h + 2h^2\right)\|\mathbf{Q}\|_2^2\right)\|\mathbf{Q}\|_2 \underbrace{\|\mathbf{G}\|_0}_{=\|I_h\mathbf{Q}-\mathbf{Q}_h\|_0},
\end{aligned}
$$

implying that

$$\|I_h\mathbf{Q} - \mathbf{Q}_h\|_0 \lesssim h^2\left(2 + \left(3 + 2h + 2h^2\right)\|\mathbf{Q}\|_2^2\right)\|\mathbf{Q}\|_2. \tag{9.26}$$

By the triangle inequality and Lemma 9.5, we have

$$\|\mathbf{Q} - \mathbf{Q}_h\|_0 \leq \|\mathbf{Q} - I_h\mathbf{Q}\|_0 + \|I_h\mathbf{Q} - \mathbf{Q}_h\|_0$$
$$\lesssim h^2 \left(2 + \left(3 + 2h + 2h^2\right) \|\mathbf{Q}\|_2^2\right) \|\mathbf{Q}\|_2.$$

This completes the proof. □

Remark 9.6. One can follow (Maity *et al.*, 2020) to obtain optimal error estimates in both norms $\|\cdot\|_1$ and $\|\cdot\|_0$ for higher degrees (≥ 2) of approximating polynomials. We omit further details since $\|\cdot\|_1$ is actually equivalent to the norm $\|\cdot\|_{dG}$ (in Maity *et al.*, 2020) in the $\mathbf{H}_0^1(\Omega)$ space and the technique can be directly applied to our case here.

In this section, we have obtained the optimal *a priori* error estimates of the regular solution $\mathbf{Q}$ in the L^2-norm (see Theorem 9.8) and in the H^1-norm (see Theorem 9.2). We will verify this in our subsequent numerical experiments in Section 9.3 with different choices of approximations.

9.2.2   *A priori error estimates for* $(\mathcal{P}2)$

Since the PDE (9.4) for the density variation u is a fourth-order problem, a conforming discretisation requires a finite-dimensional subspace of the Sobolev space $H^2(\Omega)$, which necessitates the use of $\mathcal{C}^1$-continuous elements. The construction of these elements is quite involved, particularly in three dimensions; without a special mesh structure, the lowest-degree conforming elements are the Argyris *et al.* (1968) and Zhang (2009) elements, of degree 5 and 9 in two and three dimensions, respectively. One approach to avoid this is to use mixed formulations by solving two second-order systems, and we refer to Cheng *et al.* (2000) and Scholtz (1978) for instance. However, this substantially increases the size of the linear systems to be solved. Alternatively, one can directly tackle the fourth-order problem with nonconforming elements, that do not satisfy the $\mathcal{C}^1$-requirement. For instance, the so-called *continuous/discontinuous Galerkin* (C/DG) methods and $\mathcal{C}^0$ *interior penalty* methods ($\mathcal{C}^0$-IP) are analysed in Brenner and Sung (2005) and Engel *et al.* (2002), combining concepts from the theory of continuous and discontinuous Galerkin methods. Essentially, these methods use $\mathcal{C}^0$-conforming elements and penalise inter-element jumps in first derivatives to weakly enforce $\mathcal{C}^1$-continuity. This has the advantages of both convenience and efficiency: the weak form is simple, with only minor modifications from a conforming method, and fewer degrees of freedom are used than with a fully discontinuous Galerkin method.

We thus adopt the idea of C^0-IP methods to solve the nonlinear fourth-order problem ($\mathcal{P}2$). Specifically, we use the usual C^0-continuous Lagrange elements and penalise jumps of the gradient across facets. In what follows, we derive some *a priori* error estimates for the fourth-order problem ($\mathcal{P}2$) with the strong form derived in (9.4).

For simplicity, we only consider the cubic nonlinearity (i.e., $a_2 = 0$) in this analysis. The quadratic term can be tackled similarly. We therefore analyse the following strong form:

$$\begin{cases} 2B\nabla \cdot \left(\nabla \cdot \left(\mathcal{D}^2 u\right)\right) + a_1 u + a_3 u^3 = 0 & \text{in } \Omega, \\ u = u_b & \text{on } \partial\Omega, \\ \mathcal{D}^2 u \cdot \nu = \mathcal{D}^2 u_b \cdot \nu & \text{on } \partial\Omega. \end{cases} \quad (9.27)$$

The corresponding continuous weak form is defined as: find $u \in H^2(\Omega) \cap H^1_b(\Omega)$ such that

$$\mathcal{N}^s(u)v := A^s(u,v) + B^s(u,u,u,v) + C^s(u,v) = L^s(v)$$
$$\forall v \in H^2(\Omega) \cap H^1_0(\Omega), \quad (9.28)$$

where for $v, w \in H^2(\Omega)$,

$$A^s(v,w) = 2B \int_\Omega \mathcal{D}^2 v : \mathcal{D}^2 w,$$

$$C^s(v,w) = a_1 \int_\Omega vw,$$

$$L^s(v) := 2B \int_{\partial\Omega} \left(\mathcal{D}^2 u_b \cdot \nabla v\right) \cdot \nu,$$

and for $\mu, \zeta, \eta, \xi \in H^2(\Omega)$,

$$B^s(\mu, \zeta, \eta, \xi) = a_3 \int_\Omega \mu\zeta\eta\xi.$$

Since (9.28) is nonlinear, we derive its linearisation: find $v \in H^2(\Omega) \cap H^1_0(\Omega)$ such that

$$\langle \mathcal{D}\mathcal{N}^s(u)v, w \rangle_{H^2} := A^s(v,w) + 3B^s(u,u,v,w) + C^s(v,w)$$
$$= L^s(w) \quad \forall w \in H^2(\Omega) \cap H^1_0(\Omega), \quad (9.29)$$

where $\langle \cdot, \cdot \rangle_{H^2}$ represents the dual pairing between $\left(H^2(\Omega) \cap H_0^1(\Omega)\right)^*$ and $H^2(\Omega) \cap H_0^1(\Omega)$.

It is straightforward to derive the coercivity and boundedness of the bilinear operator $A^s(\cdot, \cdot)$ with the semi-norm $|\cdot|_2$ (in fact, this is indeed a norm in $H^2(\Omega) \cap H_0^1(\Omega)$).

Lemma 9.9. *For $v, w \in H^2(\Omega) \cap H_0^1(\Omega)$, there holds*

$$A^s(v, w) \lesssim |v|_2 |w|_2 \quad and \quad A^s(v, v) \gtrsim |v|_2^2.$$

Define the broken Sobolev space by

$$H^2(\mathcal{T}_h) := \{v \in H^1(\Omega) : v|_T \in H^2(T) \; \forall T \in \mathcal{T}_h\},$$

equipped with the broken norm $\|v\|_{2,\mathcal{T}_h}^2 = \sum_{T \in \mathcal{T}_h} \|v\|_{2,T}^2$.

We take the nonconforming but still continuous approximation u_h for the solution u of (9.28), that is to say, $u_h \in W_{h,b} \subset H^2(\mathcal{T}_h) \cap H_b^1(\Omega)$ with some related definitions for deg ≥ 2

$$W_h := \{v \in H^2(\mathcal{T}_h) \cap H^1(\Omega) : v \in \mathbb{Q}_{\deg}(T) \; \forall T \in \mathcal{T}_h\},$$

$$W_{h,0} := \{v \in H^2(\mathcal{T}_h) \cap H^1(\Omega) : v = 0 \text{ on } \partial\Omega, v \in \mathbb{Q}_{\deg}(T) \; \forall T \in \mathcal{T}_h\},$$

$$W_{h,b} := \{v \in H^2(\mathcal{T}_h) \cap H^1(\Omega) : v = u_b \text{ on } \partial\Omega, v \in \mathbb{Q}_{\deg}(T) \; \forall T \in \mathcal{T}_h\}.$$

Following the derivation of $\mathcal{C}^0$-IP formulation similar to Brenner (2011, Section 3), we introduce the discrete nonlinear weak form: find $u_h \in W_{h,b}$ such that

$$\mathcal{N}_h^s(u_h)v_h := A_h^s(u_h, v_h) + P_h^s(u_h, v_h) + B_h^s(u_h, u_h, u_h, v_h) + C_h^s(u_h, v_h)$$

$$= L^s(v_h) \quad \forall v_h \in W_{h,0}, \tag{9.30}$$

where for all $u, v, \mu, \zeta, \eta, \xi \in W_h$,

$$A_h^s(u, v) := 2B \left(\sum_{T \in \mathcal{T}_h} \int_T \mathcal{D}^2 u : \mathcal{D}^2 v - \sum_{e \in \mathcal{E}_I} \int_e \left\{\!\left\{ \frac{\partial^2 u}{\partial \nu^2} \right\}\!\right\} [\![\nabla v]\!] \right.$$

$$\left. - \sum_{e \in \mathcal{E}_I} \int_e \left\{\!\left\{ \frac{\partial^2 v}{\partial \nu^2} \right\}\!\right\} [\![\nabla u]\!] \right),$$

$$C_h^s(u, v) = C^s(u, v) = a_1 \int_\Omega uv,$$

$$B_h^s(\mu, \zeta, \eta, \xi) = B^s(\mu, \zeta, \eta, \xi) = a_3 \int_\Omega \mu\zeta\eta\xi,$$

and

$$P_h^s(u, v) := \sum_{e \in \mathcal{E}_I} \frac{2B\epsilon}{h_e^3} \int_e [\![\nabla u]\!][\![\nabla v]\!]. \tag{9.31}$$

Here, ϵ is the penalty parameter (to be specified in the implementations later), the average $\left\{\!\!\left\{ \frac{\partial^2 u}{\partial \nu^2} \right\}\!\!\right\}$ of the second derivatives of u along tangential directions across e is defined as

$$\left\{\!\!\left\{ \frac{\partial^2 u}{\partial \nu^2} \right\}\!\!\right\} = \frac{1}{2} \left(\left.\frac{\partial^2 u_+}{\partial \nu^2}\right|_e + \left.\frac{\partial^2 u_-}{\partial \nu^2}\right|_e \right),$$

with ν denoting the outward normal. In fact, the operator P_h^s penalises the first derivatives across the interior facet since the function in $H^1(\Omega)$ is not necessarily continuously differentiable.

Remark 9.7. The nonlinear problems (9.28) and (9.30) are equivalent for the solution u of the strong form (9.4) since the jump term $[\![\nabla u]\!]$ vanishes for $u \in H^2(\Omega)$, however they are *not* equivalent for $u_h \in W_{h,b} \subset H^1(\Omega)$.

The linearised version of the discrete nonlinear problem (9.30) yields the following discrete linear weak form: seek $v_h \in W_{h,0}$ such that

$$\langle \mathcal{D}\mathcal{N}_h^s(u_h)v_h, w_h \rangle = L^s(w_h) \quad \forall w_h \in W_{h,0}, \tag{9.32}$$

where

$$\langle \mathcal{D}\mathcal{N}_h^s(u_h)v_h, w_h \rangle := A_h^s(v_h, w_h) + P_h^s(v_h, w_h) + 3B_h^s(u_h, u_h, v_h, w_h)$$
$$+ C_h^s(v_h, w_h). \tag{9.33}$$

We also define the mesh-dependent H^2-like semi-norm for $v \in W_h$,

$$\|\|v\|\|_h^2 := \sum_{T \in \mathcal{T}_h} |v|_{H^2(T)}^2 + \sum_{e \in \mathcal{E}_I} \int_e \frac{1}{h_e^3} |[\![\nabla v]\!]|^2. \tag{9.34}$$

Note that $\|\|\cdot\|\|_h$ is indeed a norm on $W_{h,0}$. This norm will be used in the well-posedness and convergence analysis as follows.

We first give an immediate result about the consistency of the discrete form (9.30).

Theorem 9.10 (Consistency). *Assuming that $u \in H^4(\Omega)$. The solution u of the continuous weak form (9.28) solves the discrete weak problem (9.30).*

Proof. Multiplying the fourth-order term $2B\nabla \cdot (\nabla \cdot (\mathcal{D}^2 u))$ in (9.27) with $v \in W_{h,0}$ and using piecewise integration by parts with the boundary condition specified in (9.27) for u, one can obtain

$$2B \sum_{T \in \mathcal{E}_h} \int_T \nabla \cdot (\nabla \cdot (\mathcal{D}^2 u))v = 2B \sum_{T \in \mathcal{E}_h} \int_T \mathcal{D}^2 u : \mathcal{D}^2 v$$

$$- 2B \sum_{e \in \mathcal{E}_I} \int_e \left\{\!\!\left\{ \frac{\partial^2 u}{\partial \nu^2} \right\}\!\!\right\} [\![\nabla v]\!]. \qquad (9.35)$$

Since $u \in H^4(\Omega)$ implies ∇u is continuous on the whole domain Ω, the jump term $[\![\nabla u]\!]$ then becomes zero and we can thus symmetrise and penalise the form (9.35). This leads to the presence of $A_h^s(u,v) + P_h^s(u,v)$. The remaining terms involving B_h^s and C_h^s are straightforward as one takes the test function $v \in W_{h,0}$. Therefore, u satisfies (9.30). $\qquad\square$

9.2.2.1 *Elliptic regularity*

Essentially, the strong form (9.27) is similar to the model problem given as Brenner (2011, Example 2) of form

$$\Delta^2 u = f \quad \text{in } \Omega,$$
$$u = \Delta u = 0 \quad \text{on } \partial\Omega. \qquad (9.36)$$

Remark 9.8. The boundary condition for the second derivative of u in (9.36) is different from what we have imposed in (9.27). We just want to comment about the regularity of the problem (9.27) by extending the results for (9.36).

Noticing that

$$\left(\mathcal{D}^2 : \mathcal{D}^2\right) u = \left[\left(\partial_x^2\right)^2 + \left(\partial_y^2\right)^2 + 2\left(\partial_{xy}^2\right)^2 \right] u = \Delta^2 u,$$

it is natural to extend the classical elliptic regularity result (Blum and Bonn, 1980) for the biharmonic operator Δ^2 to the case of the bi-Hessian operator $\mathcal{D}^2 : \mathcal{D}^2$. In general, the weak solution of (9.36) in a bounded polygonal domain Ω belongs to $H^{2+\varkappa}(\Omega)$ for some elliptic regularity index $\varkappa \in (0,2]$. More specifically, by Blum and Bonn (1980, Theorem 2), we know that if each interior angle is smaller than $\pi/2$, then for $f \in H^{-1}(\Omega)$

there holds

$$\|u\|_{H^3(\Omega)} \lesssim \|f\|_{H^{-1}(\Omega)}.$$

In addition, if the domain Ω is smooth, the weak solutions even belong to $H^4(\Omega)$ by classical elliptic regularity results and thus we take this as an assumption throughout the analysis for simplicity.

Hence, we assume the solution u of the strong form (9.27) is sufficiently regular in what follows and only consider to approximate such regular or nonsingular solutions of the continuous weak form (9.28). Moreover, to facilitate the following analysis, we further assume that u is an isolated solution, i.e., within a sufficiently small ball $\{v \in H^2(\Omega) \cap H_0^1(\Omega) : |v-u|_2 \leq r_b\}$ with radius r_b, there is only one solution u satisfying (9.27). These assumptions then imply that the linearised operator $\langle \mathcal{DN}^s(u)\cdot, \cdot \rangle_{H^2}$ satisfies the following inf-sup condition:

$$
\begin{aligned}
0 < \beta_u &= \inf_{\substack{v \in H^2(\Omega) \cap H_0^1(\Omega) \\ |v|_2 = 1}} \sup_{\substack{w \in H^2(\Omega) \cap H_0^1(\Omega) \\ |w|_2 = 1}} \langle \mathcal{DN}^s(u)v, w \rangle_{H^2} \\
&= \inf_{\substack{w \in H^2(\Omega) \cap H_0^1(\Omega) \\ |w|_2 = 1}} \sup_{\substack{v \in H^2(\Omega) \cap H_0^1(\Omega) \\ |v|_2 = 1}} \langle \mathcal{DN}^s(u)v, w \rangle_{H^2}.
\end{aligned}
\tag{9.37}
$$

9.2.2.2 Well-posedness of the discrete form

Recalling Brenner (2011, Eq. (3.20)) that for $u, v \in W_{h,0}$,

$$
\sum_{e \in \mathcal{E}_I} \left| \int_e \left\{\!\!\left\{ \frac{\partial^2 w}{\partial \nu^2} \right\}\!\!\right\} [\![\nabla v]\!] \right| \lesssim \left(\sum_{T \in \mathcal{T}_h} \int_T \mathcal{D}^2 w : \mathcal{D}^2 w \right)^{1/2} \\
\times \left(\sum_{e \in \mathcal{E}_I} \frac{1}{h_e} \int_e ([\![\nabla v]\!])^2 \right)^{1/2},
$$

we can immediately obtain

$$
\sum_{e \in \mathcal{E}_I} \left| \int_e \left\{\!\!\left\{ \frac{\partial^2 w}{\partial \nu^2} \right\}\!\!\right\} [\![\nabla v]\!] \right| \lesssim \left(\sum_{T \in \mathcal{T}_h} \int_T \mathcal{D}^2 w : \mathcal{D}^2 w \right)^{1/2} \\
\times \left(\sum_{e \in \mathcal{E}_I} \frac{1}{h_e^3} \int_e ([\![\nabla v]\!])^2 \right)^{1/2},
\tag{9.38}
$$

since the edge or facet size $h_e < 1$. With the estimate (9.38) at hand, we can apply the Cauchy–Schwarz inequality and use the definition (9.34) of $\||\cdot\||_h$ to obtain the boundedness of $A_h^s(\cdot, \cdot)$ and $P_h^s(\cdot, \cdot)$. That is to say, for $u, v \in W_{h,0}$, there holds

$$|A_h^s(u, v)| \lesssim \||u\||_h \||v\||_h,$$

$$|P_h^s(u, v)| \lesssim \||u\||_h \||v\||_h.$$

We omit the details of their proofs here and only illustrate the boundedness result for $B_h^s(\cdot, \cdot, \cdot, \cdot)$ and $C_h^s(\cdot, \cdot)$ below.

Lemma 9.11 (Boundedness of $B_h^s(\cdot, \cdot, \cdot, \cdot)$ and $C_h^s(\cdot, \cdot)$). *For* $u, v, w, p \in W_{h,0}$, *we have*

$$
\begin{aligned}
|B_h^s(u, v, w, p)| &\lesssim \||u\||_h \||v\||_h \||w\||_h \||p\||_h, \\
|C_h^s(u, v)| &\lesssim \||u\||_h \||v\||_h.
\end{aligned}
\tag{9.39}
$$

For $u, v \in H^2(\Omega)$, $w, p \in W_h$,

$$|B_h^s(u, v, w, p)| \lesssim \|u\|_2 \|v\|_2 \||w\||_h \||p\||_h. \tag{9.40}$$

Proof. By Hölder's inequality, Sobolev embedding $H^1(\Omega) \hookrightarrow L^4(\Omega)$, and the fact that the H^1 semi-norm $|\cdot|_1$ is a norm in $H_0^1(\Omega)$, we deduce

$$
\begin{aligned}
|B_h^s(u, v, w, p)| &\lesssim \|u\|_{L^4} \|v\|_{L^4} \|w\|_{L^4} \|p\|_{L^4} \\
&\lesssim |u|_1 |v|_1 |w|_1 |p|_1.
\end{aligned}
$$

It then follows from a Poincaré inequality (Brenner, 2003; Brenner and Sung, 2005, Eq. (4.22)) for piecewise H^1 functions that

$$\sum_{T \in \mathcal{T}_h} |v|_{1,T}^2 \lesssim \sum_{T \in \mathcal{T}_h} |v|_{2,T}^2 + \sum_{e \in \mathcal{E}_I} \frac{1}{h_e^3} \|[\![\nabla v]\!]\|_{0,e}^2 = \||v\||_h^2 \quad \forall v \in W_{h,0}. \tag{9.41}$$

Thus, we obtain

$$|B_h^s(u, v, w, p)| \lesssim \||u\||_h \||v\||_h \||w\||_h \||p\||_h.$$

The boundedness of $C_h^s(\cdot, \cdot)$ follows similarly by Cauchy–Schwarz inequality, the Sobolev embedding $H^1(\Omega) \hookrightarrow L^2(\Omega)$ and the use of (9.41).

The proof of (9.40) is analogous to that of (9.39) with a use of the embedding result $H^2(\Omega) \hookrightarrow L^\infty(\Omega)$ and the Cauchy–Schwarz inequality. $\qquad \square$

We give the coercivity result for the bilinear form $(A_h^s(\cdot,\cdot) + P_h^s(\cdot,\cdot))$.

Lemma 9.12 (Coercivity of $A_h^s + P_h^s$). *For a sufficiently large penalty parameter ϵ, there holds*

$$\|\|v_h\|\|_h^2 \lesssim A_h^s(v_h, v_h) + P_h^s(v_h, v_h) \quad \forall v_h \in W_{h,0}. \tag{9.42}$$

Proof. By (9.38) and the inequality of geometric and arithmetic means, we deduce for $v \in W_h$,

$$A_h^s(v,v) + P_h^s(v,v) \geq 2B \sum_{T \in \mathcal{T}_h} |v|_{H^2(T)}^2$$

$$- 2BC \left(\sum_{T \in \mathcal{T}_h} |v|_{2,T}^2 \right)^{1/2} \left(\sum_{e \in \mathcal{E}_I} \frac{1}{h_e^3} \|[\![\nabla v]\!]\|_{0,e}^2 \right)^{1/2}$$

$$+ 2B \left(\sum_{e \in \mathcal{E}_I} \int_e \frac{\epsilon}{h_e^3} |[\![\nabla v]\!]|^2 \right)$$

$$\geq 2B \left[\frac{1}{2} \sum_{T \in \mathcal{T}_h} |v|_{H^2(T)}^2 + \left(\epsilon - \frac{C^2}{2} \right) \sum_{e \in \mathcal{E}_I} \frac{1}{h_e^3} \|[\![\nabla v]\!]\|_{0,e}^2 \right]$$

$$\geq B \|\|v\|\|_h^2,$$

provided the penalty parameter ϵ is sufficiently large with the generic constant C from (9.38). □

An important question about the well-posedness is the coercivity of the bilinear operator $\langle \mathcal{DN}_h^s(u_h)\cdot, \cdot \rangle$. Due to the presence of B_h^s and C_h^s terms in $\langle \mathcal{DN}_h^s(u_h)\cdot, \cdot \rangle$, it is not trivial to derive its coercivity. Instead, we discuss the weak coercivity of the bilinear form $\langle \mathcal{DN}_h^s(u)\cdot, \cdot \rangle$ defined as

$$\langle \mathcal{DN}_h^s(u)v_h, w_h \rangle := A_h^s(v_h, w_h) + P_h^s(v_h, w_h)$$

$$+ 3B_h^s(u, u, v_h, w_h) + C_h^s(v_h, w_h) \quad \forall v_h, w_h \in W_h.$$

We first give a useful lemma illustrating some estimates related to the enrichment operator $E_h : W_h \to W_C \subset H^2(\Omega)$ with W_C being the Hsieh–Clough–Tocher macro finite element space. The degrees of freedom of $w \in W_C$ include: (i) the values of the derivatives of w up to order 1 at the interior vertices and (ii) the values of the normal derivative of w at the midpoints of the interior edges/facets in $\mathcal{E}_I$. The following lemma is adapted to our notations and definition of $\|\cdot\|_h$ using the result (Brenner, 2011, Lemma 1).

Lemma 9.13 (Brenner, 2011, Lemma 1). *For $v_h \in W_{h,0}$, there holds that*

$$\sum_{T \in \mathcal{T}_h} \left(h^{-4} \|v_h - E_h v_h\|_{L^2(T)}^2 + h^{-2} |v_h - E_h v_h|_{H^1(T)}^2 + |v_h - E_h v_h|_{H^2(T)}^2 \right)$$

$$\approx \sum_{e \in \mathcal{E}_I} \frac{1}{h_e^3} \|[\![\nabla v_h]\!]\|_{L^2(e)}^2 \approx \|v_h\|_h^2. \tag{9.43}$$

We can obtain the discrete inf-sup condition for the discrete bilinear operator $\langle \mathcal{D}\mathcal{N}_h^s(u)\cdot, \cdot \rangle$.

Theorem 9.14 (Weak coercivity of $\langle \mathcal{D}\mathcal{N}_h^s(u)\cdot, \cdot \rangle$). *Let u be a regular isolated solution of the nonlinear continuous weak form* (9.30). *For a sufficiently large ϵ and a sufficiently small mesh size h, the following discrete inf-sup condition holds on a smooth domain Ω with a positive constant $\beta_c > 0$:*

$$0 < \beta_c \leq \inf_{\substack{v \in W_h \\ \|v_h\|_h = 1}} \sup_{\substack{w \in W_h \\ \|w_h\|_h = 1}} \langle \mathcal{D}\mathcal{N}_h^s(u) v_h, w_h \rangle. \tag{9.44}$$

Proof. For $v \in H^2(\Omega) \cap H_0^1(\Omega)$, it follows from the boundedness result of B_h^s, C_h^s that $B_h^s(u, u, v, \cdot)$, $B^s(u, u, v, \cdot)$, $C_h^s(v, \cdot)$ and $C^s(v, \cdot) \in L^2(\Omega)$. Furthermore, since $A^s(\cdot, \cdot)$ is bounded and coercive as given by Lemma 9.9, for a given $v_h \in W_h$ with $\|v_h\|_h = 1$, there exists ξ and $\eta \in H^4(\Omega) \cap H_0^1(\Omega)$ that solve the linear systems:

$$A^s(\xi, w) = 3B_h^s(u, u, v_h, w) + C_h^s(v_h, w) \quad \forall w \in H^2(\Omega) \cap H_0^1(\Omega), \tag{9.45a}$$

$$A^s(\eta, w) = 3B^s(u, u, E_h v_h, w) + C^s(E_h v_h, w) \quad \forall w \in H^2(\Omega) \cap H_0^1(\Omega). \tag{9.45b}$$

It then follows from the standard elliptic regularity result that $\|\eta\|_4 \lesssim C_{BC}$ with constant C_{BC} depending on $\|u\|_2$.

Subtracting (9.45a) from (9.45b), then taking $w = \eta - \xi$ and using the coercivity of $A^s(\cdot, \cdot)$ and boundedness of B_h^s, C_h^s, we obtain

$$|\eta - \xi|_2 \lesssim \left(3\|u\|_2^2 + 1 \right) \|E_h v_h - v_h\|_0$$

$$\lesssim h^2 \underbrace{\|v_h\|_h}_{=1} \qquad \qquad \text{by Lemma 9.13.} \tag{9.46}$$

Here, we have used the fact that B_h^s and C_h^s are in fact equivalent to B^s and C^s, respectively, by their definitions. Since u is a regular isolated solution of (9.28), it yields by (9.37) that there exists $w \in H^2(\Omega) \cap H_0^1(\Omega)$ with $|w|_2 = 1$ such that

$$|E_h v_h|_2 \lesssim \langle \mathcal{D}\mathcal{N}^s(u)E_h v_h, w \rangle_{H^2}$$

$$= A^s(E_h v_h, w) + 3B^s(u, u, E_h v_h, w) + C^s(E_h v_h, w)$$

$$\overset{(9.45b)}{=} A^s(E_h v_h + \eta, w)$$

$$\overset{\text{Lemma 9.9}}{\lesssim} |E_h v_h + \eta|_2 \underbrace{|w|_2}_{=1}$$

$$= \||E_h v_h + \eta\||_h$$

$$\leq \||E_h v_h - v_h\||_h + \||v_h + I_h \xi\||_h + \||I_h \xi - \xi\||_h + \underbrace{\||\xi - \eta\||_h}_{=|\xi - \eta|_2}.$$
$$\tag{9.47}$$

Here, the second last equality is due to the fact that $E_h v_h + \eta \in H^2$ while the last inequality is derived from the triangle inequality. Note that $\llbracket \nabla \xi \rrbracket = 0$ on $\mathcal{E}_I$ since $\xi \in H^4(\Omega)$. We can thus calculate

$$\||E_h v_h - v_h\||_h^2 \lesssim \sum_{e \in \mathcal{E}_I} \int_e \frac{1}{h_e^3} |\llbracket \nabla v_h \rrbracket|^2 \qquad \text{by Lemma 9.13,}$$

$$\lesssim \sum_{e \in \mathcal{E}_I} \int_e \frac{1}{h_e^3} |\llbracket \nabla(v_h + \xi) \rrbracket|^2$$

$$\leq \||v_h + \xi\||_h^2.$$

Further, by the triangle inequality, we get

$$\||E_h v_h - v_h\||_h \lesssim \||v_h + \xi\||_h \leq \||v_h + I_h \xi\||_h + \||\xi - I_h \xi\||_h. \tag{9.48}$$

Since $v_h + I_h \xi \in W_h$, it follows from the coercivity result (9.42) that there exists $w_h \in W_h$ with $\||w\||_h = 1$ such that

$$\||v_h + I_h \xi\||_h \lesssim A_h^s(v_h + I_h \xi, w_h) + P_h^s(v_h + I_h \xi, w_h)$$

$$= \langle \mathcal{D}\mathcal{N}_h^s(u)v_h, w_h \rangle - 3B_h^s(u, u, v_h, w_h) - C_h^s(v_h, w_h)$$

$$\quad + A_h^s(I_h \xi - \xi, w_h) + P_h^s(I_h \xi - \xi, w_h)$$

$$\quad + A_h^s(\xi, w_h) + P_h^s(\xi, w_h)$$

$$= \langle \mathcal{D}\mathcal{N}_h^s(u)v_h, w_h \rangle + 3B_h^s(u, u, v_h, E_h w_h - w_h)$$

$$+ C_h^s(v_h, E_h w_h - w_h) + A_h^s(I_h \xi - \xi, w_h)$$

$$+ P_h^s(I_h \xi - \xi, w_h) + A_h^s(\xi, w_h - E_h w_h)$$

$$+ P_h^s(\xi, w_h - E_h w_h), \tag{9.49}$$

where in the last equality we have used the fact that

$$3B_h^s(u, u, v_h, E_h w_h) + C_h^s(v_h, E_h w_h)$$

$$= A^s(\xi, E_h w_h) = A_h^s(\xi, E_h w_h) + P^s(\xi, E_h w_h)$$

because of (9.45a) and $[\![\nabla \xi]\!] = [\![\nabla E_h w_h]\!] = 0$.

Using the boundedness result Lemma 9.11 and the enrichment estimates Lemma 9.13, we obtain

$$3B_h^s(u, u, v_h, E_h w_h - w_h) + C_h^s(v_h, E_h w_h - w_h)$$

$$\lesssim \underbrace{\|v_h\|_0}_{\lesssim |v_h|_1 \lesssim \|\|v_h\|\|_h = 1} \underbrace{\|E_h w_h - w_h\|_0}_{\lesssim h^2 \|\|w_h\|\|_h = h^2}. \tag{9.50}$$

By the boundedness of the bilinear form $A_h^s + P_h^s$ and standard interpolation estimates, we have

$$A_h^s(I_h \xi - \xi, w_h) + P_h^s(I_h \xi - \xi, w_h) \lesssim \|\|I_h \xi - \xi\|\|_h \underbrace{\|\|w_h\|\|_h}_{=1}$$

$$\lesssim h^{\min\{\deg - 1, 2\}} \|\xi\|_4, \tag{9.51}$$

where $\deg \geq 2$ denotes the degree of the approximating polynomials. Moreover, by the enrichment estimate Lemma 9.13 and the fact that $[\![\nabla \xi]\!] = [\![\nabla (E_h w_h)]\!] = 0$, there holds

$$A_h^s(\xi, w_h - E_h w_h) + P_h^s(\xi, w_h - E_h w_h)$$

$$= 2B \sum_{T \in \mathcal{T}_h} \int_T \mathcal{D}^2 \xi : \mathcal{D}^2(w_h - E_h w_h)$$

$$- 2B \sum_{e \in \mathcal{E}_I} \int_e \left\{\!\!\left\{ \frac{\partial^2 \xi}{\partial \nu^2} \right\}\!\!\right\} [\![\nabla(w_h - E_h w_h)]\!]$$

$$\overset{(9.35)}{=} 2B \sum_{T \in \mathcal{T}_h} \nabla \cdot \left(\nabla \cdot (\mathcal{D}^2 \xi) \right) (w_h - E_h w_h)$$

$$\lesssim \|\xi\|_4 \|w_h - E_h w_h\|_0$$

$$\overset{\text{Lemma 9.13}}{\lesssim} h^2 \|\xi\|_4 \tag{9.52}$$

Combine Equations (9.50) to (9.52) in (9.49) to obtain

$$\|\!|\!| E_h v_h - v_h |\!|\!\|_h \lesssim \langle \mathcal{D} \mathcal{N}_h^s(u) v_h, w_h \rangle + h^2 + h^{\min\{\deg -1, 2\}}. \tag{9.53}$$

Substituting (9.53) into (9.48) and using standard interpolation estimates yield that

$$\|\!|\!| E_h v_h - v_h |\!|\!\|_h \lesssim \langle \mathcal{D} \mathcal{N}_h^s(u) v_h, w_h \rangle + h^2 + h^{\min\{\deg -1, 2\}}. \tag{9.54}$$

A use of Equations (9.53) and (9.54), standard interpolation estimates and (9.46) in (9.47) leads to

$$|E_h v_h|_2 \lesssim \langle \mathcal{D} \mathcal{N}_h^s(u) v_h, w_h \rangle + h^2 + h^{\min\{\deg -1, 2\}}.$$

Then, by the triangle inequality, we have

$$1 = \|\!|\!| v_h |\!|\!\|_h \leq \|\!|\!| v_h - E_h v_h |\!|\!\|_h + \underbrace{\|\!|\!| E_h v_h |\!|\!\|_h}_{=|E_h v_h|_2}$$

$$\leq C_t \left(\langle \mathcal{D} \mathcal{N}_h^s(u) v_h, w_h \rangle + h^2 + h^{\min\{\deg -1, 2\}} \right).$$

Therefore, for the mesh size h satisfying

$$h^2 + h^{\min\{\deg -1, 2\}} < \frac{1}{2 C_t},$$

the discrete inf-sup condition (9.44) holds for $\beta_c = \frac{1}{2 C_t}$. $\qquad \square$

Moreover, we can obtain the discrete inf-sup condition for the perturbed bilinear form $\langle \mathcal{D} \mathcal{N}_h^s(I_h u) \cdot, \cdot \rangle$, i.e.,

$$\langle \mathcal{D} \mathcal{N}_h^s(I_h u) v_h, w_h \rangle = A_h^s(v_h, w_h) + P_h^s(v_h, w_h)$$

$$+ 3 B_h^s(I_h u, I_h u, v_h, w_h) + C_h^s(v_h, w_h) \quad \forall v_h, w_h \in W_h. \tag{9.55}$$

Theorem 9.15 (Weak coercivity of $\langle \mathcal{D} \mathcal{N}_h^s(I_h u) \cdot, \cdot \rangle$). *Let u be a regular isolated solution of the nonlinear continuous weak form* (9.30) *and $I_h u$*

the interpolation of u. For a sufficiently large ϵ and a sufficiently small mesh size h, the following discrete inf-sup condition holds:

$$0 < \frac{\beta_c}{2} \le \inf_{\substack{v_h \in W_h \\ \|v_h\|_h = 1}} \sup_{\substack{w_h \in W_h \\ \|w_h\|_h = 1}} \langle \mathcal{DN}_h^s(I_h u) v_h, w_h \rangle. \tag{9.56}$$

Proof. Denote $\tilde{u} = u - I_h u$. By the definition (9.55) of the bilinear form $\langle \mathcal{DN}_h^s(I_h u)\cdot, \cdot \rangle$, we have

$$\langle \mathcal{DN}_h^s(I_h u) v_h, w_h \rangle = A_h^s(v_h, w_h) + P_h^s(v_h, w_h) + 3 B_h^s(u - \tilde{u}, u - \tilde{u}, v_h, w_h)$$
$$+ C_h^s(v_h, w_h).$$

It follows from the definition of B_h^s and its boundedness result Lemma 9.11 that

$$B_h^s(u - \tilde{u}, u - \tilde{u}, v_h, w_h)$$
$$= B_h^s(u, u, v_h, w_h) + B_h^s(\tilde{u}, \tilde{u}, v_h, w_h) - 2 B_h^s(u, \tilde{u}, v_h, w_h)$$
$$\ge B_h^s(u, u, v_h, w_h) + B_h^s(\tilde{u}, \tilde{u}, v_h, w_h) - 2 C_1 \|u\|_h \|\tilde{u}\|_h \|v_h\|_h \|w_h\|_h,$$

where C_1 is the generic constant arising in the boundedness result Lemma 9.11 for $B_h^s(\cdot, \cdot, \cdot, \cdot)$. Therefore, we obtain that

$$\langle \mathcal{DN}_h^s(I_h u) v_h, w_h \rangle$$
$$\ge \langle \mathcal{DN}_h^s(u) v_h, w_h \rangle + 3 B_h^s(\tilde{u}, \tilde{u}, v_h, w_h) - 6 C_1 \|u\|_h \|\tilde{u}\|_h \|v_h\|_h \|w_h\|_h.$$

Now using the inf-sup condition Theorem 9.14 for the bilinear form $\langle \mathcal{DN}_h^s(u)\cdot, \cdot \rangle$, boundedness result Lemma 9.11 and interpolation estimates, we get

$$\sup_{\substack{\|w_h\|_h = 1 \\ w_h \in W_h}} \langle \mathcal{DN}_h^s(I_h u) v_h, w_h \rangle \ge \sup_{\substack{\|w_h\|_h = 1 \\ w_h \in W_h}} \langle \mathcal{DN}_h^s(u) v_h, w_h \rangle - 3 |B_h^s(\tilde{u}, \tilde{u}, v_h, w_h)|$$

$$- 6 C_1 h^{\min\{\deg - 1, \Bbbk_u - 2\}} \|u\|_h \|v_h\|_h$$

$$\ge \left(\beta_c - C_2 h^{\min\{\deg - 1, \Bbbk_u - 2\}} \right) \|v_h\|_h$$

$$\ge \frac{\beta_c}{2} \|v_h\|_h,$$

for a sufficiently small mesh size h such that $h^{\min\{\deg - 1, \Bbbk_u - 2\}} < \frac{\beta_c}{2 C_2}$. Here, C_2 depends on C_1 and $\|u\|_2$ and $\Bbbk_u > 2$ gives the regularity of u, i.e., $u \in H^{\Bbbk_u}(\Omega)$. Therefore, the inf-sup condition (9.56) holds. $\qquad \square$

9.2.2.3 *Convergence analysis*

We proceed to the error analysis for the discrete nonlinear problem (9.30). Let

$$\mathcal{B}_\rho(I_h u) := \{v_h \in W_h : \|I_h u - v_h\|_h \leq \rho\},$$

where I_h is the interpolation operator mapping from the infinite-dimensional space $H^2(\mathcal{T}_h) \cap H^1(\Omega)$ to the finite-dimensional space W_h. We define the nonlinear map $\mu_h : W_h \to W_h$ by

$$\langle \mathcal{DN}_h^s(I_h u_h)\mu_h(v_h), w_h \rangle = 3B_h^s(I_h u_h, I_h u_h, v_h, w_h) + L^s(w_h)$$
$$- B_h^s(v_h, v_h, v_h, w_h). \tag{9.57}$$

Due to the weak coercivity property in Theorem 9.15, the nonlinear map μ_h is well-defined.

The existence and local uniqueness result of the discrete solution u_h to the discrete nonlinear problem (9.30) will be proven via an application of Brouwer's fixed point theorem, which necessitates the use of two auxiliary lemmas illustrating that (i) μ_h maps from a ball to itself; and (ii) the map μ_h is contracting.

Lemma 9.16 (Mapping from a ball to itself). *Let u be a regular isolated solution of the continuous nonlinear weak problem (9.28). For a sufficiently large ϵ and a sufficiently small mesh size h, there exists a positive constant $R(h) > 0$ such that:*

$$\|v_h - I_h u\|_h \leq R(h) \Rightarrow \|\mu_h(v_h) - I_h u\|_h \leq R(h) \quad \forall v_h \in W_{h,0}.$$

Proof. Note that the solution $u \in H^2(\Omega) \cap H_0^1(\Omega)$ of (9.28) satisfies the discrete weak formulation (9.30) due to the consistency result Theorem 9.10, that is to say, there holds that

$$A_h^s(u, w_h) + P_h^s(u, w_h) + B_h^s(u, u, u, w_h)$$
$$+ C_h^s(u, w_h) = L^s(w_h) \quad \forall w_h \in W_{h,0}. \tag{9.58}$$

By the linearity of $\langle \mathcal{DN}_h^s(I_h u)\cdot, \cdot \rangle_{H^2}$, the definition (9.57) of the nonlinear map μ_h and formulation (9.58), we calculate

$$\langle \mathcal{DN}_h^s(I_h u)(I_h u - \mu_h(v_h)), w_h \rangle$$
$$= \langle \mathcal{DN}_h^s(I_h u)I_h u, w_h \rangle - \langle \mathcal{DN}_h^s(I_h u)\mu_h(v_h), w_h \rangle$$
$$= A_h^s(I_h u, w_h) + P_h^s(I_h u, w_h) + 3B_h^s(I_h u, I_h u, I_h u, w_h) + C_h^s(I_h u, w_h)$$

$$-3B_h^s(I_h u, I_h u, v_h, w_h) + B_h^s(v_h, v_h, v_h, w_h) - L^s(w_h)$$

$$= \underbrace{A_h^s(I_h u - u, w_h) + P_h^s(I_h u - u, w_h)}_{\mathfrak{N}_1} + \underbrace{C_h^s(I_h u - u, w_h)}_{\mathfrak{N}_2}$$

$$+ \underbrace{(B_h^s(I_h u, I_h u, I_h u, w_h) - B_h^s(u, u, u, w_h))}_{\mathfrak{N}_3}$$

$$+ \underbrace{(2B_h^s(I_h u, I_h u, I_h u, w_h) - 3B_h^s(I_h u, I_h u, v_h, w_h) + B_h^s(v_h, v_h, v_h, w_h))}_{\mathfrak{N}_4}$$

$$=: \mathfrak{N}_1 + \mathfrak{N}_2 + \mathfrak{N}_3 + \mathfrak{N}_4.$$

In what follows, we give the upper bounds for each $\mathfrak{N}_i$, $i = 1, 2, 3, 4$. A use of the boundedness of $A_h^s + P_h^s, C_h^s$ and the interpolation estimate (Brenner and Sung, 2005, Eq. (5.3)) in the $\|\|\cdot\|\|$-norm, we obtain

$$\mathfrak{N}_1 \lesssim \|\|I_h u - u\|\|_h \|\|w_h\|\|_h \lesssim h^{\min\{\deg-1, \Bbbk_u - 2\}} \|\|w_h\|\|_h,$$

$$\mathfrak{N}_2 \lesssim \|\|I_h u - u\|\|_h \|\|w_h\|\|_h \lesssim h^{\min\{\deg-1, \Bbbk_u - 2\}} \|\|w_h\|\|_h.$$

We rearrange terms in $\mathfrak{N}_3$ and use the boundedness result Lemma 9.11 and the interpolation result (Brenner and Sung, 2005, Eq. (5.3)) to obtain

$$\mathfrak{N}_3 = B_h^s(I_h u, I_h u, I_h u, w_h) - B_h^s(u, u, u, w_h)$$

$$= B_h^s(I_h u - u, I_h u - u, I_h u, w_h)$$

$$\quad + 2B_h^s(I_h u - u, I_h u - u, u, w_h) + 3B_h^s(u, u, I_h u - u, w_h)$$

$$\lesssim \left(\|I_h u - u\|_h^2 \|I_h u\|_h + \|I_h u - u\|_h^2 \|u\|_h + \|u\|_2^2 \|I_h u - u\|_0 \right) \|\|w_h\|\|_h$$

$$\lesssim \left(h^{2\min\{\deg-1, \Bbbk_u - 2\}} + h^{\min\{\deg+1, \Bbbk_u\}} \right) \|\|w_h\|\|_h.$$

Let $e_I = v_h - I_h u$. We use the definition of $B_h^s(\cdot, \cdot, \cdot, \cdot)$ and use its boundedness to deduce that

$$\mathfrak{N}_4 = 2B_h^s(I_h u, I_h u, I_h u, w_h) - 3B_h^s(I_h u, I_h u, v_h, w_h) + B_h^s(v_h, v_h, v_h, w_h)$$

$$= a_3 \int_\Omega \left\{ 2(I_h u)^3 w_h - 3(I_h u)^2 v_h w_h + v_h^3 w_h \right\}$$

$$= a_3 \int_\Omega \left\{ \left(v_h^2 - (I_h u)^2 \right) v_h w_h + 2(I_h u)^2 (I_h u - v_h) w_h \right\}$$

$$= a_3 \int_\Omega \left\{ e_I(e_I + 2I_h u)(e_I + I_h u)w_h - 2(I_h u)^2 e_I w_h \right\}$$

$$= a_3 \int_\Omega \left\{ e_I \left(e_I^2 + 3e_I I_h u + 2(I_h u)^2 \right) w_h - 2(I_h u)^2 e_I w_h \right\}$$

$$= a_3 \int_\Omega \left(e_I^3 + 3e_I^2 I_h u \right) w_h$$

$$= B_h^s(e_I, e_I, e_I, w_h) + 3B_h^s(e_I, e_I, I_h u, w_h)$$

$$\lesssim \|e_I\|_h^2 \left(\|e_I\|_h + \|I_h u\|_h \right) \|w_h\|_h.$$

Hence, we combine the above bounds for $\mathfrak{N}_i$, $i = 1, 2, 3, 4$ to have

$$\langle D\mathcal{N}_h^s(I_h u)(I_h u - \mu_h(v_h)), w_h \rangle$$

$$\lesssim \left(h^{\min\{\deg-1, \mathbb{k}_u-2\}} + h^{\min\{2\deg-2, 2\mathbb{k}_u-4, \deg+1, \mathbb{k}_u\}} \right.$$

$$\left. + \|e_I\|_h^2 \left(\|e_I\|_h + 1 \right) \right) \|w_h\|_h.$$

By the inf-sup condition (9.56) for the perturbed bilinear form, we further deduce that there exists a $w_h \in W_h$ with $\|w_h\|_h = 1$ such that

$$\|I_h u - \mu_h(v_h)\|_h \lesssim \langle D\mathcal{N}_h^s(I_h u)(I_h u - \mu_h(v_h)), w_h \rangle.$$

Since $\|e_I\|_h \leq R(h)$, we obtain

$$\|I_h u - \mu_h(v_h)\|_h$$

$$\lesssim \left(h^{\min\{\deg-1, \mathbb{k}_u-2\}} + h^{\min\{2\deg-2, 2\mathbb{k}_u-4, \deg+1, \mathbb{k}_u\}} + R(h)^2 \left(R(h) + 1 \right) \right)$$

$$\leq \begin{cases} C_u \left(2h^{\min\{\deg-1, \mathbb{k}_u-2\}} + R(h)^2(1 + R(h)) \right), \\ \qquad \text{for } 2 \leq \deg \leq 3, \mathbb{k}_u \leq 4, \\ C_u \left(h^{\min\{\deg-1, \mathbb{k}_u-2\}} + h^{\min\{\deg+1, 2\mathbb{k}_u-4\}} + R(h)^2(1 + R(h)) \right), \\ \qquad \text{for } \deg > 3, \mathbb{k}_u \leq 4. \end{cases}$$

Note that there are other cases when $\mathbb{k}_u > 4$ and we only focus on the case of $\mathbb{k}_u \leq 4$ here for brevity. Hence, the idea of the remainder of the proof is to choose an appropriate $R(h)$ so that $\|I_h u - \mu_h(v_h)\|_h \leq R(h)$. For simplicity of the calculation, we illustrate the case when $2 \leq \deg \leq 3$, $\mathbb{k}_u \leq 4$. To this end, we take $R(h) = 4C_u h^{\min\{\deg-1, \mathbb{k}_u-2\}}$ and choose h

satisfying

$$h^{2\min\{\deg-1,\Bbbk_u-2\}} \leq \frac{1}{32C_u} - \frac{1}{16}.$$

This yields

$$\||I_h u - \mu_h(v_h)\||_h \leq 2C_u h^{\min\{\deg-1,\Bbbk_u-2\}} \left(1 + C_u R(h)^2 + C_u\right)$$

$$= 2C_u h^{\min\{\deg-1,\Bbbk_u-2\}} \left(1 + 32C_u^3 h^{2\min\{\deg-1,\Bbbk_u-2\}} + 2C_u\right)$$

$$\leq R(h).$$

This completes the proof. $\square$

Lemma 9.17 (Contraction result). *For a sufficiently large ϵ, a sufficiently small mesh size h and any $v_1, v_2 \in \mathcal{B}_{R(h)}(I_h u)$, there holds*

$$\||\mu_h(v_1) - \mu_h(v_2)\||_h \lesssim h^{\min\{\deg-1,\Bbbk_u-2\}} \||v_1 - v_2\||_h. \tag{9.59}$$

Proof. For $w_h \in W_h$, we use the definition (9.57) of the nonlinear map μ_h, definition and linearity of $\langle \mathcal{D}\mathcal{N}_h^s(I_h u)\cdot, \cdot\rangle$ to calculate

$$\langle \mathcal{D}\mathcal{N}_h^s(I_h u)(\mu_h(v_1) - \mu_h(v_2)), w_h\rangle$$

$$= 3B_h^s(I_h u, I_h u, v_1, w_h) - B_h^s(v_1, v_1, v_1, w_h)$$

$$\quad - 3B_h^s(I_h u, I_h u, v_2, w_h) + B_h^s(v_2, v_2, v_2, w_h)$$

$$= a_3 \int_\Omega \left(3(I_h u)^2 v_1 w_h - v_1^3 w_h\right) - a_3 \int_\Omega \left(3(I_h u)^2 v_2 w_h - v_2^3 w_h\right)$$

$$= a_3 \int_\Omega \left(\left((I_h u)^2 - v_1^2\right) v_1 w_h + 2(I_h u)^2(v_1 - v_2)w_h - \left((I_h u)^2 - v_2^2\right) v_2 w_h\right)$$

$$= a_3 \int_\Omega ((I_h u - v_1)(v_1 - I_h u)(v_1 - v_2)w_h + 2(I_h u - v_1)I_h u(v_1 - v_2)w_h$$

$$\quad + (I_h u - v_1)(I_h u + v_1)v_2 w_h)$$

$$\quad + 2a_3 \int_\Omega (I_h u(v_1 - v_2)(I_h u - v_2)w_h + I_h u(v_1 - v_2)v_2 w_h)$$

$$\quad - a_3 \int_\Omega (I_h u - v_2)(I_h u + v_2)v_2 w_h$$

$$= a_3 \int_\Omega (I_h u - v_1)(v_1 - I_h u)(v_1 - v_2) w_h$$

$$+ 2a_3 \int_\Omega (I_h u - v_1) I_h u (v_1 - v_2) w_h$$

$$+ 2a_3 \int_\Omega (I_h u - v_2) I_h u (v_1 - v_2) w_h$$

$$+ a_3 \int_\Omega (v_1 - v_2) \left((I_h u - v_1) + (I_h u - v_2) \right) \left((v_2 - I_h u) + I_h u \right) w_h.$$

Let $e_1 = I_h u - v_1$, $e_2 = I_h u - v_2$ and $e = v_1 - v_2$. We make some elementary manipulations and use the boundedness result of B_h^s and the inequality of geometric and arithmetic means to get

$$\langle \mathcal{D} \mathcal{N}_h^s (I_h u)(\mu_h(v_1) - \mu_h(v_2)), w_h \rangle$$

$$= a_3 \int_\Omega (-e_1^2) e w_h + 2a_3 \int_\Omega e_1 (I_h u) e w_h + 2a_3 \int_\Omega e_2 (I_h u) e w_h$$

$$+ a_3 \int_\Omega \{ e w_h (e_1 I_h u + e_2 I_h u - e_1 e_2 - e_2^2) \}$$

$$\lesssim \left(\| e_1 \|_h^2 + \| I_h u \|_h \| e_1 \|_h + \| e_2 \|_h \| I_h u \|_h + \| e_1 \|_h \| e_2 \|_h + \| e_2 \|_h^2 \right)$$

$$\times \| e \|_h \| w_h \|_h$$

$$\lesssim \left(\| e_1 \|_h^2 + \| e_2 \|_h^2 + \| e_1 \|_h + \| e_2 \|_h \right) \| e \|_h \| w_h \|_h$$

$$\lesssim \left(R(h)^2 + R(h) \right) \| e \|_h \| w_h \|_h.$$

By the inf-sup condition (9.56), we know that there exist $w_h \in W_h$ with $\| w_h \|_h = 1$, such that

$$\frac{\beta_c}{2} \| \mu_h(v_1) - \mu_h(v_2) \|_h \lesssim \langle \mathcal{D} \mathcal{N}_h^s (I_h u)(\mu_h(v_1) - \mu_h(v_2)), w_h \rangle.$$

Therefore, we have

$$\| \mu_h(v_1) - \mu_h(v_2) \|_h \lesssim R(h)(1 + R(h)) \| e \|_h.$$

Note that $R(h)(1 + R(h)) < 1$ for $R(h) < 1$. This completes the proof. $\square$

The existence and local uniqueness of the discrete solution u_h can now be obtained via the application of Brouwer's fixed point theorem (Kesavan, 1989).

Theorem 9.18 (Convergence in $\|\|\cdot\|\|_h$-norm). *Let u be a regular isolated solution of the nonlinear problem* (9.28). *For a sufficiently large ϵ and a sufficiently small h, there exists a unique solution u_h of the discrete nonlinear problem* (9.30) *within the local ball $\mathcal{B}_{R(h)}(I_h u)$. Furthermore, we have the following bound:*

$$\|\|u - u_h\|\|_h \lesssim h^{\min\{\deg -1, \Bbbk_u - 2\}},$$

where $\deg \geq 1$ *denotes the degree of the polynomial approximation and* $\Bbbk_u \geq 2$ *is the regularity index of u.*

Proof. A use of Lemma 9.16 yields that the nonlinear map μ_h maps a closed convex set $\mathcal{B}_{R(h)}(I_h u) \subset W_h$ to itself. Moreover it is a contracting map. Therefore, an application of the Brouwer fixed point theorem (Kesavan, 1989) yields that μ_h has at least one fixed point, say u_h, in this ball $\mathcal{B}_{R(h)}(I_h u)$. The uniqueness of the solution to (9.30) in that ball $\mathcal{B}_{R(h)}(I_h u)$ follows from the contraction result in Lemma 9.17. Meanwhile, we have by Lemma 9.16 that

$$\|\|u_h - I_h u\|\|_h \lesssim h^{\min\{\deg -1, \Bbbk_u - 2\}}. \tag{9.60}$$

The error estimate is then obtained straightforwardly using the triangle inequality

$$\|\|u - u_h\|\|_h \leq \|\|u - I_h u\|\|_h + \|\|I_h u - u_h\|\|_h,$$

combined with (9.60) and the interpolation estimate (Brenner and Sung, 2005, Eq. (5.3)). $\Box$

It is implied from Theorem 9.18 that optimal convergence rates have been shown in the mesh-dependent norm $\|\|\cdot\|\|_h$. We will see the numerical verifications of this in Section 9.3.

9.2.2.4 *Estimates in the L^2-norm*

We derive an L^2 error estimate using a duality argument in this section. To this end, we consider the following linear dual problem to the primary nonlinear problem (9.4):

$$\begin{cases} 2B\nabla \cdot (\nabla \cdot (\mathcal{D}^2 \chi)) + a_1 \chi + 3a_3 u^2 \chi = f_{\text{dual}} & \text{in } \Omega, \\ \chi = 0 & \text{on } \partial\Omega, \\ \nu \cdot \mathcal{D}^2 \chi = \mathbf{0} & \text{on } \partial\Omega, \end{cases} \tag{9.61}$$

for $f_{\text{dual}} \in L^2(\Omega)$. For smooth domains Ω, it can be deduced by a classical elliptic regularity result that $\chi \in H^4(\Omega)$. The corresponding weak form is derived: find $\chi \in H^2(\Omega) \cap H^1_0(\Omega)$ such that

$$2B \int_\Omega \mathcal{D}^2\chi : \mathcal{D}^2 v + a_1 \int_\Omega \chi v + 3a_3 \int_\Omega u^2\chi v$$

$$= \int_\Omega f_{\text{dual}} v \quad \forall v \in H^2(\Omega) \cap H^1_0(\Omega),$$

that is to say,

$$\langle \mathcal{D}\mathcal{N}^s(u)\chi, v \rangle_{H^2} = \langle \mathcal{D}\mathcal{N}^s_h(u)\chi, v \rangle = (f_{\text{dual}}, v)_0. \tag{9.62}$$

Remark 9.9. The first equality in (9.62) holds since $u \in H^2(\Omega), \chi \in H^2(\Omega)$ and $v \in H^2(\Omega)$.

We give two auxiliary results in the following.

Lemma 9.19. *For $u \in H^{\Bbbk_u}(\Omega)$, $\Bbbk_u > 2$, $\chi \in H^4(\Omega) \cap H^1_0(\Omega)$ and $I_h u \in W_{h,0} \subset H^1_0(\Omega)$, there holds that*

$$A^s_h(I_h u - u, \chi) + P^s_h(I_h u - u, \chi) \lesssim h^{\min\{\deg+1,\Bbbk_u\}}\|\chi\|_4.$$

Proof. Note that $[\![\nabla\chi]\!] = 0$ since $\chi \in H^4(\Omega)$ and $\chi = 0$ on $\partial\Omega$. We calculate

$$A^s(I_h u - u, \chi) + P^s_h(I_h u - u, \chi)$$

$$= \sum_{T \in \mathcal{E}_h} \int_T 2B\mathcal{D}^2(I_h u - u) : \mathcal{D}^2\chi$$

$$- 2B \sum_{e \in \mathcal{E}_I} \left\{\!\!\left\{ \frac{\partial^2(I_h u - u)}{\partial\nu^2} \right\}\!\!\right\} [\![\nabla\chi]\!] - 2B \sum_{e \in \mathcal{E}_I} \left\{\!\!\left\{ \frac{\partial^2\chi}{\partial\nu^2} \right\}\!\!\right\} [\![\nabla(I_h u - u)]\!]$$

$$+ \sum_{e \in \mathcal{E}_I} \frac{2B\epsilon}{h_e^3} \int_e [\![\nabla(I_h u - u)]\!][\![\nabla\chi]\!]$$

$$= \sum_{T \in \mathcal{E}_h} \int_T 2B\mathcal{D}^2(I_h u - u) : \mathcal{D}^2\chi - 2B \sum_{e \in \mathcal{E}_I} \left\{\!\!\left\{ \frac{\partial^2\chi}{\partial\nu^2} \right\}\!\!\right\} [\![\nabla(I_h u - u)]\!]$$

$$= \sum_{T \in \mathcal{E}_h} \int_T 2B(I_h u - u)\nabla \cdot (\nabla \cdot (\mathcal{D}^2\chi))$$

$$\lesssim \|I_h u - u\|_0 \|\nabla \cdot (\nabla \cdot (\mathcal{D}^2\chi))\|_0$$

$$\lesssim h^{\min\{\deg+1,\Bbbk_u\}}\|\chi\|_4.$$

Here, the last, second last, third last steps follow from the standard interpolation estimates, the Cauchy–Schwarz inequality, and integration by parts twice, respectively. □

Lemma 9.20. *The solution χ of the linear dual problem* (9.61) *belongs to* $H^4(\Omega)$ *on a smooth domain* Ω *and it holds that*

$$\|\chi\|_4 \lesssim \|f_{\mathrm{dual}}\|_0. \tag{9.63}$$

Proof. We can use the inf-sup condition (9.37) for the linear operator $\langle \mathcal{DN}^s(u)\cdot, \cdot \rangle$, the weak form (9.62) and the Cauchy–Schwarz inequality to obtain

$$|\chi|_2 \lesssim \sup_{\substack{w \in H^2 \cap H_0^1 \\ |w|_2=1}} \langle \mathcal{DN}^s(u)\chi, w \rangle_{H^2} = \sup_{\substack{w \in H^2 \cap H_0^1 \\ |w|_2=1}} (f_{\mathrm{dual}}, w)_0 \lesssim \|f\|_0 \underbrace{\|w\|_0}_{\lesssim |w|_2=1}. \tag{9.64}$$

By the form of (9.62) and the boundedness of $B^s(u, u, \cdot, \cdot)$ and $C^s(\cdot, \cdot)$, we have

$$\|2B\nabla \cdot (\nabla \cdot (\mathcal{D}^2\chi))\|_0 = \| -3B^s(u, u, \chi, \cdot) - C^s(\chi, \cdot) + (f_{\mathrm{dual}}, \cdot)_0\|_0$$

$$\lesssim \underbrace{\|\chi\|_0}_{\lesssim |\chi|_2} + \|f_{\mathrm{dual}}\|_0$$

$$\overset{(9.64)}{\lesssim} \|f_{\mathrm{dual}}\|_0. \tag{9.65}$$

Using a bootstrapping argument in elliptic regularity (see, e.g., Evans, 2010, Section 6.3), we can deduce that $\chi \in H^4(\Omega)$ in a smooth domain Ω. Moreover, it is implied from (9.65) that the regularity estimate (9.63) holds. □

We are ready to derive the L^2 *a priori* error estimates.

Theorem 9.21 (L^2 error estimate). *Under the same conditions as in Theorem 9.18 and assuming further that* $\deg \geq 1, \Bbbk_u \geq 2$, *the discrete solution* u_h *approximates* u *such that*

$$\|u - u_h\|_0 \lesssim \begin{cases} h^{\min\{\deg+1, \Bbbk_u\}} & \text{for } \deg \geq 3, \\ h^{2\min\{\deg-1, \Bbbk_u-2\}} & \text{for } \deg = 2. \end{cases}$$

Proof. Taking $f_{\text{dual}} = I_h u - u_h \in W_h \subset H^1(\Omega) \cap H^2(\mathcal{T}_h)$ in (9.61) and multiplying (9.61) by a test function $v_h = I_h u - u_h$ with integration by parts, we obtain

$$\langle \mathcal{D}\mathcal{N}_h^s(u)\chi, I_h u - u_h \rangle = \|I_h u - u_h\|_0^2.$$

It follows from the fact that $u \in H^{\Bbbk_u}(\Omega)$, $\Bbbk_u \geq 2$, and the definition (9.28) of the nonlinear continuous weak form $\mathcal{N}^s(u)$· that

$$
\begin{aligned}
&\|I_h u - u_h\|_0^2 \\
&= \langle \mathcal{D}\mathcal{N}_h^s(u)\chi, I_h u - u_h \rangle + \mathcal{N}_h^s(u_h)(I_h\chi) - \mathcal{N}_h^s(u)(I_h\chi) \\
&= A_h^s(\chi, I_h u - u_h) + P_h^s(\chi, I_h u - u_h) + C_h^s(\chi, I_h u - u_h) + 3B_h^s(u, u, \chi, I_h u - u_h) \\
&\quad + A_h^s(u_h, I_h\chi) + P_h^s(u_h, I_h\chi) + C_h^s(u_h, I_h\chi) + B_h^s(u_h, u_h, u_h, I_h\chi) \\
&\quad - A_h^s(u, I_h\chi) - P_h^s(u, I_h\chi) - C_h^s(u, I_h\chi) - B_h^s(u, u, u, I_h\chi) \\
&= \underbrace{A_h^s(I_h u - u, \chi) + A_h^s(u - u_h, \chi - I_h\chi) + P_h^s(I_h u - u, \chi) + P_h^s(u - u_h, \chi - I_h\chi)}_{\mathfrak{U}_1} \\
&\quad + \underbrace{C_h^s(I_h u - u, \chi) + C_h^s(u - u_h, \chi - I_h\chi)}_{\mathfrak{U}_2} \\
&\quad + \underbrace{3B_h^s(u, u, I_h u - u_h, \chi - I_h\chi) + 3B_h^s(u, u, I_h u - u, I_h\chi)}_{\mathfrak{U}_3} \\
&\quad + \underbrace{B_h^s(u_h, u_h, u_h, I_h\chi) - 3B_h^s(u, u, u_h, I_h\chi) + 2B_h^s(u, u, u, I_h\chi)}_{\mathfrak{U}_4} \\
&=: \mathfrak{U}_1 + \mathfrak{U}_2 + \mathfrak{U}_3 + \mathfrak{U}_4.
\end{aligned}
$$

We then bound each $\mathfrak{U}_i$ separately using the boundedness results for A_h^s, P_h^s, B_h^s and C_h^s and standard interpolation estimates. This leads to

$$\mathfrak{U}_1 \lesssim h^{\min\{\deg+1,\Bbbk_u\}} \|\chi\|_4 + \underbrace{\|\|u - u_h\|\|_h}_{\lesssim h^{\min\{\deg-1,\Bbbk_u-2\}}} \underbrace{\|\|\chi - I_h\chi\|\|_h}_{\lesssim h^2\|\chi\|_4} \quad \text{by Theorem 9.18,}$$

$$\lesssim h^{\min\{\deg+1,\Bbbk_u\}} \|\chi\|_4,$$

$$\mathfrak{U}_2 \lesssim \underbrace{\|I_h u - u\|_0}_{\lesssim h^{\min\{\deg+1,\Bbbk_u\}}} \underbrace{\|\chi\|_0}_{\leq \|\chi\|_4} + \|\|u - u_h\|\|_h \|\|\chi - I_h\chi\|\|_h$$

$$\lesssim h^{\min\{\deg+1,\Bbbk_u\}} \|\chi\|_4,$$

and

$$\mathfrak{U}_3 = 3B_h^s(u, u, I_h u - u_h, \chi - I_h \chi) + 3B_h^s(u, u, I_h u - u, I_h \chi)$$
$$\lesssim \|u\|_2^2 \underbrace{\|\|I_h u - u_h\|\|_h}_{\lesssim h^{\min\{\deg - 1, \Bbbk_u - 2\}}} \underbrace{\|\|\chi - I_h \chi\|\|_h}_{\lesssim h^2 \|\chi\|_4} + \|u\|_2^2 \underbrace{\|I_h u - u\|_0}_{\lesssim h^{\min\{\deg + 1, \Bbbk_u\}}} \underbrace{\|I_h \chi\|_0}_{\lesssim \|\chi\|_4}$$
$$\lesssim h^{\min\{\deg + 1, \Bbbk_u\}} \|\chi\|_4.$$

Setting $e_3 = u_h - u$ and estimating $\mathfrak{U}_4$ as in $\mathfrak{R}_4$ of Lemma 9.16 yield

$$\mathfrak{U}_4 \lesssim \|\|e_3\|\|_h^2 (\|\|e_3\|\|_h + \|\|u\|\|_h) \underbrace{\|\|I_h \chi\|\|_h}_{\lesssim \|\chi\|_2 \lesssim \|\chi\|_4}$$
$$\lesssim h^{2\min\{\deg - 1, \Bbbk_u - 2\}} (h^{\min\{\deg - 1, \Bbbk_u - 2\}} + 1) \|\chi\|_4 \quad \text{by Theorem 9.18.}$$

Combining the above estimates for $\mathfrak{U}_i$ $(i = 1, 2, 3, 4)$ and using the regularity estimate (9.63), we obtain

$$\|I_h u - u_h\|_0^2 \lesssim \begin{cases} h^{\min\{\deg + 1, \Bbbk_u\}} \underbrace{\|\chi\|_4}_{\lesssim \|I_h u - u_h\|_0} & \text{if } \deg \geq 3, \Bbbk_u \geq 4, \\ h^{2\min\{\deg - 1, \Bbbk_u - 2\}} \underbrace{\|\chi\|_4}_{\lesssim \|I_h u - u_h\|_0} & \text{if } \deg = 2, \Bbbk_h \leq 4. \end{cases}$$

Using the triangle inequality and standard interpolation estimates, we get

$$\|u - u_h\|_0 \leq \|u - I_h u\|_0 + \|I_h u - u_h\|_0$$
$$\lesssim h^{\min\{\deg + 1, \Bbbk_u\}} + \begin{cases} h^{\min\{\deg + 1, \Bbbk_u\}} & \text{for } \deg \geq 3, \\ h^{2\min\{\deg - 1, \Bbbk_u - 2\}} & \text{for } \deg = 2, \end{cases}$$
$$\lesssim \begin{cases} h^{\min\{\deg + 1, \Bbbk_u\}} & \text{for } \deg \geq 3, \\ h^{2\min\{\deg - 1, \Bbbk_u - 2\}} & \text{for } \deg = 2. \end{cases}$$

This complies the proof. $\qquad\square$

Theorem 9.21 implies that for quadratic approximations to the sufficiently regular solution of (9.27), there is a sub-optimal convergence rate in the L^2-norm while for higher order (≥ 3) approximations, we expect optimal L^2 error rates. We shall see numerical verifications of this in the subsequent sections.

9.2.2.5   *The inconsistent discrete form*

The above analysis considers the consistent weak formulation (9.30) in finite element discretisations. In practice, we adopt the inconsistent discrete weak form in the implementations in Chapter 10 because of its simplicity in the discrete weak form: find $u_h \in W_{h,b}$ such that for all $v_h \in W_{h,0}$,

$$\tilde{\mathcal{N}}_h^s(u_h)v_h = \tilde{A}_h^s(u_h, v_h) + B_h^s(u_h, u_h, u_h, v_h) + C_h^s(u_h, v_h) + P_h^s(u_h, v_h) = 0,$$
$$(9.66)$$

where

$$\tilde{A}_h^s(u, v) := 2B \sum_{T \in \mathcal{T}_h} \int_T \mathcal{D}^2 u : \mathcal{D}^2 v.$$

Note that the missing terms by comparing $\tilde{A}_h^s$ and A_h^s are those interelement summations involving the average of the second tangential derivatives, arising from piecewise integration by parts and the symmetrisation. Due to the absence of those terms in $\tilde{A}_h^s$, one can immediately notice that the discrete weak formulation (9.66) is inconsistent in the sense that the solution u of the strong form (9.27) does not satisfy the weak form (9.66), as opposed to Theorem 9.10.

Regardless of this inconsistency that complicates the convergence analysis, our choice of the discrete weak form (9.30) reduces the complexity of the implementation and in practice leads to a converging numerical scheme (though may not possess optimal convergence rates), as illustrated in Section 9.3. This is not surprising; a similar idea has also been applied and introduced as *weakly over-penalised symmetric interior penalty* (WOPSIP) methods in Brenner and Sung (2008) for second-order elliptic PDEs and in Brenner *et al.* (2010) for biharmonic equations.

Remark 9.10. The excessive size of the penalty parameter in the WOPSIP method could induce ill-conditioned linear systems. It is also discussed in Brenner and Sung (2008) how to design block preconditioners and analyse the conditioning of the linear systems. Moreover, in all of our numerical experiments in the next section, we do not observe any ill-conditioning effects.

In our numerical examinations of the convergence rate for the inconsistent discrete weak form (9.66), we find that the inconsistency does not substantially alter the convergence rate proved for the consistent form. Thus, the inconsistent formulation (9.66) can be a viable choice in implementations.

9.3   Convergence Tests

The proceeding section presents some *a priori* error estimates for the continuous Lagrange finite elements for both $\mathbf{Q}$ and u in the decoupled case $q = 0$. We now test the convergence rate of the finite element approximations by the method of manufactured solutions (MMS) and experimentally investigate the coupled case $q \neq 0$ in two dimensions. To this end, we choose a nontrivial solution for each state variable and add an appropriate source term to the equilibrium equations (see Appendix A.2 for its derivation), thus modifying the energy accordingly. We can then compute the numerical convergence order. Therefore, our chosen solution should solve the equilibrium equations exactly when we take a suitable initial guess and we can compute the numerical convergence order.

9.3.1   *Test 1: On the unit square*

In this test, the numerical runs are performed on the unit square $\Omega = (0,1)^2$ and we take the following exact expressions for each state variable:

$$Q_{11}^e = \left(\cos \left(\frac{\pi(2y-1)(2x-1)}{8} \right) \right)^2 - \frac{1}{2},$$

$$Q_{12}^e = \cos \left(\frac{\pi(2y-1)(2x-1)}{8} \right) \sin \left(\frac{\pi(2y-1)(2x-1)}{8} \right), \tag{9.67}$$

$$u^e = 10 \left((x-1)x(y-1)y \right)^3,$$

and substitute them into the derived equilibrium equations (A.5) to obtain the source terms $\mathfrak{s}_1, \mathfrak{s}_2$ and $\mathfrak{s}_3$, yielding

$$\mathfrak{s}_1 := 4Bq^4(u^e)^2 q_1^e + 2Bq^2 u^e \left(\partial_x^2 u^e - \partial_y^2 u^e \right) - 2K\Delta Q_{11}^e$$
$$- 4lQ_{11}^e + 16lq_1^e \left((Q_{11}^e)^2 + (q_2^e)^2 \right),$$

$$\mathfrak{s}_2 := 4Bq^4(u^e)^2 q_2^e + 4Bq^2 u^e \left(\partial_x \partial_y u^e \right) - 2K\Delta Q_{12}^e$$
$$- 4lQ_{12}^e + 16lq_2^e \left((Q_{11}^e)^2 + (Q_{12}^e)^2 \right),$$

$$\mathfrak{s}_3 := a_1 u^e + a_2(u^e)^2 + a_3(u^e)^3 + 2B\Delta^2 u^e$$
$$+ Bq^4 \left(4 \left((Q_{11}^e)^2 + (Q_{12}^e)^2 \right) + 1 \right) u^e + 2Bq^2(t_1^e + t_2^e),$$

with

$$t_1^e := (Q_{11}^e + 1/2)\partial_x^2 u^e + (-Q_{11}^e + 1/2)\partial_y^2 u^e + 2Q_{12}^e \partial_x \partial_y u^e,$$

$$t_2^e := \partial_x^2 (u^e (Q_{11}^e + 1/2)) + \partial_y^2 (u^e (-Q_{11}^e + 1/2)) + 2\partial_x \partial_y (u^e Q_{12}^e).$$

We take t_1 and t_2 when replacing the exact expressions of Q_{11}^e, Q_{12}^e, u^e by the unknowns Q_{11}, Q_{12}, u.

Therefore, in conducting the MMS, we are to solve the following governing equations:

$$
\begin{aligned}
&4Bq^4 u^2 Q_{11} + 2Bq^2 u \left(\partial_x^2 u - \partial_y^2 u \right) \\
&\quad - 2K\Delta Q_{11} - 4lQ_{11} + 16lQ_{11} \left(Q_{11}^2 + Q_{12}^2 \right) = \mathfrak{s}_1, \\
&4Bq^4 u^2 Q_{12} + 4Bq^2 u \left(\partial_x \partial_y u \right) \\
&\quad - 2K\Delta Q_{12} - 4lQ_{12} + 16lQ_{12} \left(Q_{11}^2 + Q_{12}^2 \right) = \mathfrak{s}_2, \\
&a_1 u + a_2 u^2 + a_3 u^3 + 2B\nabla \cdot (\nabla \cdot (\mathcal{D}^2 u)) \\
&\quad + Bq^4 \left(4 \left(Q_{11}^2 + Q_{12}^2 \right) + 1 \right) u + 2Bq^2 (t_1 + t_2) = \mathfrak{s}_3,
\end{aligned}
\tag{9.68}
$$

subject to Dirichlet boundary conditions for both u and Q and a natural boundary condition for u arising from the manufactured solutions (9.67).

We partition the domain into $N \times N$ small squares with the uniform mesh size $h = \frac{1}{N}$ ($N = 6, 12, 24, 48$) and denote numerical solutions u_h, $Q_{11,h}$ and $Q_{12,h}$. The numerical errors of u and $\mathbf{Q}$ in the $\| \cdot \|_0$-, $\| \cdot \|_1$- and $\|\|\cdot\|\|_h$-norms are defined as

$$\|\mathbf{e}_u\|_0 = \|u^e - u_h\|_0, \quad \|\mathbf{e}_u\|_1 = \|u^e - u_h\|_1, \quad \|\|\mathbf{e}_u\|\|_h = \|\|u^e - u_h\|\|_h,$$

$$\|\mathbf{e}_\mathbf{Q}\|_0 = \|(Q_{11}^e, Q_{12}^e) - (Q_{11,h}, Q_{12,h})\|_0,$$

$$\|\mathbf{e}_\mathbf{Q}\|_1 = \|(Q_{11}^e, Q_{12}^e) - (Q_{11,h}, Q_{12,h})\|_1.$$

The convergence order is then calculated from the formula

$$\log_2 \left(\frac{\text{error}_{h/2}}{\text{error}_h} \right).$$

Throughout this section, we use the parameter values

$$a_1 = -10, \ a_2 = 0, \ a_3 = 10, \ B = 10^{-5}, \ K = 0.3 \text{ and } l = 30,$$

yielding a similar choice as in the simulations of oily streaks in Section 10.4.

Remark 9.11. Since this is purely a numerical verification exercise, the manufactured solution can be physically unrealistic. Moreover, we must specify a reasonable initial guess for Newton's iteration due to the nonlinearity of the problem. The initial guess throughout this section is taken to be $\left(\frac{1}{2}(\text{exact solution}) + 10^{-9}\right)$.

9.3.1.1   *Convergence rate for $q = 0$*

In the case of $q = 0$, we essentially solve two independent nonlinear problems: one second order PDE for the tensor order parameter $\mathbf{Q}$ and a fourth order PDE for the density variation u. Therefore, we present the convergence results for $\mathbf{Q}$ and u separately in this section to verify the *a priori* error estimates proven in Section 9.2.

For the tensor variable $\mathbf{Q}$, we expect both optimal H^1 and L^2 rates, as illustrated in Theorems 9.2 and 9.8. Table 9.1 presents the numerical convergence rate for the finite elements $[\mathbb{Q}_1]^2$, $[\mathbb{Q}_2]^2$ and $[\mathbb{Q}_3]^2$. It is clear to see that optimal L^2 and H^1 rates are shown with all choices of finite elements. More specifically, second order in L^2 and first order in H^1 are observed for the approximation $[\mathbb{Q}_1]^2$. This is consistent with the proven error estimates in Section 9.2.1.

Regarding the density variation u, we first present the convergence behaviour of the consistent discrete formulation (9.30) with penalty parameter $\epsilon = 1$, since we have proven the optimal error rate in the mesh-dependent norm $\|\cdot\|_h$. The errors and convergence orders are listed

Table 9.1. Test 1: Convergence rates for $\mathbf{Q}$ with different degrees of polynomial approximation, in the decoupled case $q = 0$.

	$N = \frac{1}{h}$	$\|e_{\mathbf{Q}}\|_0$	Rate	$\|e_{\mathbf{Q}}\|_1$	Rate
$[\mathbb{Q}_1]^2$	6	8.12×10^{-4}	—	3.78×10^{-2}	—
	12	2.02×10^{-4}	2.01	1.88×10^{-2}	1.01
	24	5.05×10^{-5}	2.00	9.39×10^{-3}	1.00
	48	1.26×10^{-5}	2.00	4.69×10^{-3}	1.00
$[\mathbb{Q}_2]^2$	6	2.92×10^{-5}	—	1.11×10^{-3}	—
	12	3.90×10^{-6}	2.90	2.71×10^{-4}	2.04
	24	5.02×10^{-7}	2.96	6.72×10^{-5}	2.01
	48	6.36×10^{-8}	2.99	1.68×10^{-5}	2.00
$[\mathbb{Q}_3]^2$	6	3.02×10^{-7}	—	2.25×10^{-5}	—
	12	2.17×10^{-8}	3.80	2.72×10^{-6}	3.05
	24	1.45×10^{-9}	3.90	3.34×10^{-7}	3.03
	48	9.33×10^{-11}	3.96	4.13×10^{-8}	3.01

in Table 9.2. Optimal rates are observed in the $\|\|\cdot\|\|_h$-norm. Furthermore, optimal orders of convergence in the L^2-norm are shown for approximating polynomials of degree greater than 2, while a sub-optimal rate in the L^2-norm is given for piecewise quadratic polynomials, exactly as expected. The theoretical *a priori* error estimates are indeed verified. Sub-optimal convergence rates for quadratic polynomials were also illustrated in the numerical results of Süli and Mozolevski (2007). We also tested the convergence with the penalty parameter $\epsilon = 5 \times 10^4$ and found that the discrete norms are very similar to Table 9.2. We therefore avoid repeating the details here.

We next give the error rates for the inconsistent discrete formulation (9.66) which is actually used in the applications in Chapter 10. Though the analysis is not given for such discretisation, we wish to demonstrate that it is still convergent. We illustrate the discrete norms and the computed convergence rates in Table 9.3 with the penalty parameter $\epsilon = 1$. It can be observed that only first order of convergence is obtained in the H^2-like norm $\|\|\cdot\|\|_h$ even with different approximating polynomials. Moreover, we notice by comparing Tables 9.2 and 9.3 that the convergence rate deteriorates slightly for polynomials of degree 3 (although not for degree 4). This, however, can be improved by choosing a larger penalty parameter, as shown in Table 9.4 with $\epsilon = 5 \times 10^4$, where optimal rates are shown for the discrete norms $\|\|\cdot\|\|_h$, $\|\cdot\|_1$ and $\|\cdot\|_0$ for all polynomial degrees (except only sub-optimal in $\|\cdot\|_0$ when a piecewise quadratic polynomial is used as the approximation).

Table 9.2. Test 1: Convergence rates using the consistent discrete formulation (9.30) with penalty parameter $\epsilon = 1$ and different polynomial degrees, in the decoupled case $q = 0$.

	$N = \frac{1}{h}$	$\|e_u\|_0$	Rate	$\|e_u\|_1$	Rate	$\|\|e_u\|\|_h$	Rate
$\mathbb{Q}_2$	6	1.17×10^{-5}	—	3.46×10^{-4}	—	1.36×10^{-2}	—
	12	2.60×10^{-6}	2.17	9.81×10^{-5}	1.82	7.25×10^{-3}	0.91
	24	6.37×10^{-7}	2.03	2.54×10^{-5}	1.95	3.54×10^{-3}	1.03
	48	1.82×10^{-7}	1.80	6.88×10^{-6}	1.88	1.76×10^{-3}	1.01
$\mathbb{Q}_3$	6	4.73×10^{-6}	—	1.32×10^{-4}	—	4.98×10^{-3}	—
	12	3.32×10^{-7}	3.83	1.41×10^{-5}	3.23	9.96×10^{-4}	2.32
	24	2.12×10^{-8}	3.97	1.63×10^{-6}	3.12	2.46×10^{-4}	2.02
	48	1.32×10^{-9}	4.00	1.99×10^{-7}	3.03	6.14×10^{-5}	2.00
$\mathbb{Q}_4$	6	2.01×10^{-7}	—	7.76×10^{-6}	—	3.94×10^{-4}	—
	12	5.40×10^{-9}	5.22	4.30×10^{-7}	4.17	4.88×10^{-5}	3.01
	24	1.68×10^{-10}	5.00	2.68×10^{-8}	4.00	6.11×10^{-6}	2.99
	48	5.27×10^{-12}	4.99	1.68×10^{-9}	3.99	7.64×10^{-7}	3.00

Table 9.3. Test 1: Convergence rates using the inconsistent discrete formulation (9.66) with penalty parameter $\epsilon = 1$ and different polynomial degrees, in the decoupled case $q = 0$.

	$N = \frac{1}{h}$	$\|e_u\|_0$	Rate	$\|e_u\|_1$	Rate	$\|\|e_u\|\|_h$	Rate
$\mathbb{Q}_2$	6	3.50×10^{-6}	—	1.06×10^{-4}	—	5.60×10^{-3}	—
	12	8.76×10^{-8}	5.32	5.41×10^{-6}	4.29	2.56×10^{-3}	1.13
	24	1.77×10^{-8}	2.31	7.47×10^{-7}	2.86	1.28×10^{-3}	0.99
	48	4.35×10^{-9}	2.02	1.24×10^{-7}	2.56	6.42×10^{-4}	1.00
$\mathbb{Q}_3$	6	6.47×10^{-6}	—	1.86×10^{-4}	—	7.59×10^{-3}	—
	12	3.40×10^{-7}	4.25	1.73×10^{-5}	3.43	2.74×10^{-3}	1.47
	24	1.98×10^{-8}	4.10	2.03×10^{-6}	3.09	1.31×10^{-3}	1.07
	48	3.73×10^{-9}	2.39	2.63×10^{-7}	2.95	6.45×10^{-4}	1.02
$\mathbb{Q}_4$	6	2.05×10^{-7}	—	7.85×10^{-6}	—	3.93×10^{-4}	—
	12	5.40×10^{-9}	5.24	4.31×10^{-7}	4.19	4.88×10^{-5}	3.01
	24	1.68×10^{-10}	5.00	2.68×10^{-8}	4.01	6.11×10^{-6}	3.00
	48	5.27×10^{-12}	5.00	1.67×10^{-9}	4.00	7.64×10^{-7}	3.00

Table 9.4. Test 1: Convergence rates using the inconsistent discrete formulation (9.66) with penalty parameter $\epsilon = 5 \times 10^4$ and different polynomial degrees, in the decoupled case $q = 0$.

	$N = \frac{1}{h}$	$\|e_u\|_0$	Rate	$\|e_u\|_1$	Rate	$\|\|e_u\|\|_h$	Rate
$\mathbb{Q}_2$	6	1.17×10^{-5}	—	3.48×10^{-4}	—	1.36×10^{-2}	—
	12	2.62×10^{-6}	2.16	9.86×10^{-5}	1.82	7.26×10^{-3}	0.91
	24	6.38×10^{-7}	2.04	2.54×10^{-5}	1.96	3.54×10^{-3}	1.03
	48	1.82×10^{-7}	1.81	6.88×10^{-6}	1.88	1.76×10^{-3}	1.01
$\mathbb{Q}_3$	6	4.80×10^{-6}	—	1.35×10^{-4}	—	4.92×10^{-3}	—
	12	3.35×10^{-7}	3.84	1.43×10^{-5}	3.23	9.86×10^{-4}	2.32
	24	2.14×10^{-8}	3.97	1.63×10^{-6}	3.13	2.45×10^{-4}	2.01
	48	1.33×10^{-9}	4.01	1.99×10^{-7}	3.04	6.13×10^{-5}	2.00
$\mathbb{Q}_4$	6	2.05×10^{-7}	—	7.85×10^{-6}	—	3.93×10^{-4}	—
	12	5.40×10^{-9}	5.24	4.31×10^{-7}	4.19	4.88×10^{-5}	3.01
	24	1.68×10^{-10}	5.00	2.68×10^{-8}	4.01	6.11×10^{-6}	3.00
	48	5.27×10^{-12}	5.00	1.67×10^{-9}	4.00	7.64×10^{-7}	3.00

9.3.1.2   *Convergence rate for $q \neq 0$*

We next investigate the numerical convergence behaviour in the coupled case, i.e., $q \neq 0$, in this subsection. Its analysis is left for future work, but since it is the coupled case that is solved in practice it is important to assure ourselves that the discretisation is sensible. For brevity, we fix the model parameter $q = 30$.

We directly examine the inconsistent discretisation for u with the penalty parameter $\epsilon = 5 \times 10^4$ in the coupled case where $q \neq 0$ and fixing the $[\mathbb{Q}_2]^2$-approximation for $\mathbf{Q}$. In some unreported preliminary experiments, we observed that varying the approximations for u does not affect the convergence behaviour of $\mathbf{Q}$, that is to say, the error in $\mathbf{Q}$ depends mainly on the element used for $\mathbf{Q}$, but the polynomial that approximates u should have at least the same degree as that for $\mathbf{Q}$. We thus give the convergence rates separately for u and $\mathbf{Q}$ in Tables 9.5 and 9.6. It can be seen that $\mathbf{Q}$ retains optimal rates in both the H^1 and L^2 norms, and though there are

Table 9.5. Test 1: Convergence rates for u with $q = 30$ and penalty parameter $\epsilon = 5 \times 10^4$ with the inconsistent discretisation (9.66) for u, fixing the approximation for $\mathbf{Q}$ to be with the $[\mathbb{Q}_2]^2$ element.

	$N = \frac{1}{h}$	$\|\mathbf{e}_u\|_0$	Rate	$\|\mathbf{e}_u\|_1$	Rate	$\|\|\mathbf{e}_u\|\|_h$	Rate
$\mathbb{Q}_2$	6	1.21×10^{-5}	—	3.59×10^{-4}	—	1.37×10^{-2}	—
	12	3.98×10^{-6}	1.61	1.42×10^{-4}	1.34	8.30×10^{-3}	0.72
	24	1.57×10^{-6}	1.35	4.99×10^{-5}	1.51	3.89×10^{-3}	1.09
	48	2.58×10^{-7}	2.60	9.06×10^{-6}	2.46	1.78×10^{-3}	1.13
$\mathbb{Q}_3$	6	7.36×10^{-6}	—	2.25×10^{-4}	—	9.10×10^{-3}	—
	12	4.13×10^{-7}	4.16	1.86×10^{-5}	3.60	1.11×10^{-3}	3.03
	24	4.23×10^{-8}	3.29	2.24×10^{-6}	3.05	2.53×10^{-4}	2.14
	48	3.01×10^{-9}	3.81	2.28×10^{-7}	3.29	6.15×10^{-5}	2.04

Table 9.6. Test 1: Convergence rates for $\mathbf{Q}$ with $q = 30$ and penalty parameter $\epsilon = 5 \times 10^4$ with the inconsistent discretisation (9.66) for u, fixing the approximation for u to be with the $\mathbb{Q}_3$ element.

	$N = \frac{1}{h}$	$\|\mathbf{e_Q}\|_0$	Rate	$\|\mathbf{e_Q}\|_1$	Rate
$[\mathbb{Q}_1]^2$	6	8.12×10^{-4}	—	3.78×10^{-2}	—
	12	2.02×10^{-4}	2.01	1.88×10^{-2}	1.01
	24	5.05×10^{-5}	2.00	9.39×10^{-3}	1.00
	48	1.26×10^{-5}	2.00	4.69×10^{-3}	1.00
$[\mathbb{Q}_2]^2$	6	2.92×10^{-5}	—	1.11×10^{-3}	—
	12	3.90×10^{-6}	2.90	2.71×10^{-4}	2.04
	24	5.02×10^{-7}	2.96	6.72×10^{-5}	2.01
	48	6.37×10^{-8}	2.98	1.68×10^{-5}	2.00
$[\mathbb{Q}_3]^2$	6	3.02×10^{-7}	—	2.25×10^{-5}	—
	12	2.17×10^{-8}	3.80	2.72×10^{-6}	3.05
	24	1.45×10^{-9}	3.90	3.34×10^{-7}	3.03
	48	9.32×10^{-11}	3.96	4.13×10^{-08}	3.01

some fluctuations of the order for u, it still possesses very similar convergence rates when compared with the decoupled case described in Table 9.4.

Remark 9.12. We also tested the convergence with the consistent weak formulation for u under the same numerical settings as in Tables 9.5 and 9.6. We found that in both cases they present very similar convergence behaviour and thus we skip the details here.

Since the error norms for the finite element pair $\mathbb{Q}_3 \times [\mathbb{Q}_2]^2$ for $(u, \mathbf{Q})$ are in a rather close level of magnitude with a reasonable computational cost, we choose this approximation in our subsequent numerical experiments in Chapter 10.

Remark 9.13. We also tested the convergence with the consistent weak formulation for u under the same numerical settings as in Tables 9.5 and 9.6. We found that in both cases they present very similar convergence behaviour and thus we omit the details here.

9.3.2   *Test 2: On the unit disc*

For the second set of experiments, we provide numerical results for an exact solution u^e with only H^3-regularity, instead of C^∞ as in (9.67). Our goal is to investigate whether the H^4-regularity assumption on u can be relaxed. To this end, we consider a triangular mesh of $\Omega = \{(x, y) \mid x^2 + y^2 < 1\}$ and choose u^e to be

$$u^e = (x^2 + y^2)^{3/2}, \tag{9.69}$$

and choose the same exact solution for Q^e_{11} and Q^e_{12} as in (9.67). The exact solution given by (9.69) is in $H^3(\Omega)$ but not in $H^4(\Omega)$, hence violating the regularity assumption of the analysis in Section 9.2.2.

The resulting convergence rates are reported in Tables 9.7 and 9.8. Table 9.7 shows that optimal H^1 and L^2 rates are achieved for $\mathbf{Q}$ with three different choices of finite elements $[\mathbb{P}_1]^2, [\mathbb{P}_2]^2, [\mathbb{P}_3]^2$. Table 9.8 shows the convergence behaviour with penalty parameter $\varepsilon = 1$ when using the inconsistent discrete formulation (9.66). In contrast to Table 9.3, only first order convergence is obtained for the discrete norm $\|\!|\!|\cdot|\!|\!\|_h$ and second-order convergence for $\|\cdot\|_0$ and $\|\cdot\|_1$, with both $\mathbb{P}_3$ and $\mathbb{P}_4$. Interestingly, Table 9.8 indicates no convergence when using $\mathbb{P}_2$ elements. It appears that the assumption $u \in H^4(\Omega)$ is necessary for our analysis, and that a different analysis should be carried out when this assumption no longer holds.

Table 9.7. Test 2: Convergence rates for $\mathbf{Q}$ with different degrees of polynomial approximation, in the decoupled case $q = 0$.

	No. of triangles	$\|e_{\mathbf{Q}}\|_0$	Rate	$\|e_{\mathbf{Q}}\|_1$	Rate
$[\mathbb{P}_1]^2$	60	6.08×10^{-2}	—	1.09	—
	240	1.56×10^{-2}	1.96	5.80×10^{-1}	0.91
	960	3.92×10^{-3}	2.00	2.93×10^{-1}	0.99
	3840	9.83×10^{-4}	2.00	1.47×10^{-1}	1.00
	15360	2.47×10^{-4}	1.99	7.34×10^{-2}	1.00
$[\mathbb{P}_2]^2$	60	8.97×10^{-3}	—	2.11×10^{-1}	—
	240	1.51×10^{-3}	2.57	5.87×10^{-2}	1.84
	960	2.22×10^{-4}	2.77	1.52×10^{-2}	1.95
	3840	3.02×10^{-5}	2.88	3.85×10^{-3}	1.98
	15360	3.93×10^{-6}	2.94	9.67×10^{-4}	1.99
$[\mathbb{P}_3]^2$	60	1.08×10^{-3}	—	3.21×10^{-2}	—
	240	8.21×10^{-5}	3.72	4.58×10^{-3}	2.81
	960	5.52×10^{-6}	3.89	5.92×10^{-4}	2.95
	3840	3.54×10^{-7}	3.96	7.44×10^{-5}	2.99
	15360	2.23×10^{-8}	3.99	9.31×10^{-6}	3.00

Table 9.8. Test 2: Convergence rates using the inconsistent discrete formulation (9.66) with penalty parameter $\epsilon = 1$ and different polynomial degrees, in the decoupled case $q = 0$.

	No. of triangles	$\|e_u\|_0$	Rate	$\|e_u\|_1$	Rate	$\|e_u\|_h$	Rate
$\mathbb{P}_2$	60	1.36×10^{-2}	—	2.77×10^{-1}	—	7.39	—
	240	1.58×10^{-3}	3.11	3.86×10^{-2}	2.84	2.39	1.63
	960	7.76×10^{-4}	1.02	1.57×10^{-2}	1.30	1.44	0.73
	3840	1.84×10^{-3}	−1.25	3.42×10^{-2}	−1.12	1.26	0.19
	15360	2.77×10^{-3}	−0.59	5.29×10^{-2}	−0.63	1.60	−0.34
$\mathbb{P}_3$	60	9.94×10^{-3}	—	2.21×10^{-1}	—	6.15	—
	240	3.99×10^{-3}	1.32	8.49×10^{-2}	1.38	3.03	1.02
	960	1.33×10^{-4}	4.90	4.57×10^{-3}	4.22	1.27	1.26
	3840	3.21×10^{-5}	2.06	8.47×10^{-4}	2.43	0.66	0.93
	15360	9.22×10^{-6}	1.80	2.11×10^{-4}	2.00	0.34	0.97
$\mathbb{P}_4$	60	7.17×10^{-3}	—	1.65×10^{-1}	—	4.70	—
	240	1.34×10^{-3}	2.42	3.84×10^{-2}	2.10	2.39	0.98
	960	1.18×10^{-4}	3.50	4.54×10^{-3}	3.08	1.28	0.90
	3840	2.98×10^{-5}	1.99	8.14×10^{-4}	2.48	0.67	0.94
	15360	8.15×10^{-6}	1.87	1.86×10^{-4}	2.13	0.34	0.97

9.4   Summary

In this chapter, we derived some *a priori* error estimates related to our proposed model (8.10) for smectics and examined the convergence rates in two dimensions via the method of manufactured solutions. We focused the analysis on the decoupled case for simplicity. Optimal rates in both L^2 and H^1 norms were shown and verified for the tensor $\mathbf{Q}$. Moreover, we proved optimal convergence rates for u in the mesh-dependent norm $\|\|\cdot\|\|_h$ and the L^2 norm $\|\cdot\|_0$ (only suboptimal for piecewise quadratic polynomials). This was also illustrated in numerical experiments. By studying the convergence behaviour of different finite element choices, we noted that $\hat{\mathbb{Q}}_3 \times [\mathbb{Q}_2]^2$ for $(u, \mathbf{Q})$ with the penalty parameter $\epsilon = 5 \times 10^4$ is a suitable choice to be applied to further scenarios where physically realistic defects need to be characterised. We will apply our model and discretisation to situations of physical interest in the next chapter.

Chapter 10

Numerical Experiments for Smectics

With the convergent finite element pair $\mathbb{Q}_3 \times [\mathbb{Q}_2]^2$ for $(u, \mathbf{Q})$ at hand, we now consider three scenarios of physical interest: the defect-free example from the work of Williams and Kléman (1975), a focal conic domain simulation, and an oily streaks simulation. The first scenario is a simple example intended to examine the bending effect in smectics, while the latter two experiments depict two typical defects in smectics, thus elucidating the effectiveness of our proposed model.

For the choice of parameters, we mainly use the values suggested in Pevnyi *et al.* (2014), occasionally varying them based on physical intuition (e.g., choosing a larger wave number q to achieve thinner layers, or a larger anchoring weight w to more strongly enforce the boundary conditions). The new parameters that do not appear in the model of Pevnyi *et al.* (e.g., l and w) were chosen via unreported initial numerical experiments.

10.1 Implementation Details

As discussed in Section 9.3, we choose C^0-continuous finite element pairs for $(u, \mathbf{Q})$ with the penalty parameter $\epsilon = 5 \times 10^4$ throughout this chapter. In two dimensions, we use quadrilateral meshes. Since we restrict $\mathbf{Q}$ to be a symmetric and traceless tensor, it has two independent components in two dimensions and we thus seek the components of $\mathbf{Q}$ in $[\mathbb{Q}_2]^2$ and u in $\mathbb{Q}_3$. We utilise hexahedral meshes in three dimensions, and since $\mathbf{Q}$ has five independent components, we then seek its components in $[\mathbb{Q}_2]^5$, while retaining u in $\mathbb{Q}_3$.

In the numerical experiments, the nonlinear solve is deemed to have converged when the Euclidean norm of the residual falls below 10^{-8}, or reduces from its initial value by a factor of 10^{-8}, whichever comes first. For the inner solves, the linearised systems are solved using the sparse LU factorisation library MUMPS (Amestoy *et al.*, 2000). The mesh scale, h_e, employed in the $\mathcal{C}^0$ interior penalty approach is chosen to be the average of the diameters of the cells on either side of an edge.

To compute the stability of each solution profile, we calculate the inertia of the Hessian matrix of the energy functional with a Cholesky factorisation, implemented in MUMPS (Amestoy *et al.*, 2000). If the Hessian matrix is positive semi-definite, we characterise the solution as stable, while any nonzero number of negative eigenvalues characterises an unstable solution (Nocedal and Wright, 1999). Note that no zero eigenvalues of Hessians were observed in this chapter, i.e., the stable solutions all in fact had positive-definite Hessian matrices. For a handful of parameter values where deflation yields a solution of lowest energy that is unstable (i.e., does not find a candidate ground state), we then calculate the eigen-directions of negative curvature using the Krylov–Schur algorithm (Stewart, 2002) implemented in SLEPc (Hernandez *et al.*, 2005). We then perturb the lowest-energy solution along its eigen-directions of negative curvature and employ the bounded Newton line search algorithm of TAO (Dener *et al.*, 2020) to converge to a stable solution of minimal energy.

We give further details for the configuration of each example in the remainder of this chapter.

Code availability. For reproducibility, both the solver code (Xia, 2021b) and the exact version of Firedrake (Firedrake-Zenodo, 2021b) used to produce the numerical results in this chapter have been archived on Zenodo. An installation of Firedrake with components matching those used here can be obtained by following the instructions at https://www.firedrakeproject .org/download.html with

```
python3 firedrake-install --doi 10.5281/zenodo.4441123
```

Defcon version #11e883c should then be installed, as described in https:/ /bitbucket.org/pefarrell/defcon/.

10.2 Scenario I: Defect Free

This is a simple example proposed by the work of Williams and Kléman (1975) to examine the bending effect in smectics. For a rectangle

$\Omega = [-2, 2] \times [0, 2]$ with boundary labels

$$\Gamma_l = \{(x, y) : x = -2\}, \quad \Gamma_r = \{(x, y) : x = 2\},$$
$$\Gamma_b = \{(x, y) : y = 0\}, \quad \Gamma_t = \{(x, y) : y = 2\},$$

we strongly impose

$$\mathbf{Q} = \begin{bmatrix} (\cos\theta_0)^2 - \frac{1}{2} & -\cos\theta_0 \sin\theta_0 \\ -\cos\theta_0 \sin\theta_0 & (\sin\theta_0)^2 - \frac{1}{2} \end{bmatrix} \quad \text{on } \Gamma_b,$$

$$\mathbf{Q} = \begin{bmatrix} (\cos\theta_0)^2 - \frac{1}{2} & \cos\theta_0 \sin\theta_0 \\ \cos\theta_0 \sin\theta_0 & (\sin\theta_0)^2 - \frac{1}{2} \end{bmatrix} \quad \text{on } \Gamma_t,$$

and enforce periodic boundary conditions on the left and right boundaries, Γ_l and Γ_r. The above Dirichlet data for $\mathbf{Q}$ is derived from imposing $\mathbf{n}_e = (\cos\theta_0, -\sin\theta_0)$ at the bottom boundary, Γ_b, and with $\mathbf{n}_e = (\cos\theta_0, \sin\theta_0)$ at the top boundary, Γ_t, for fixed $\theta_0 \in [0, \pi/2]$.

We discretise the domain Ω into 90×30 quadrilateral elements and take the following initial guesses for u and $\mathbf{Q}$:

$$u = 1, \quad \mathbf{Q} = \mathbf{Q}_0, \tag{10.1}$$

where $\mathbf{Q}_0 = (\mathbf{n}_I \otimes \mathbf{n}_I - \frac{\mathbf{I}_2}{2})$ with

$$\mathbf{n}_I = \frac{1}{m_I} \begin{bmatrix} x(|x| - R) \\ (|x|)y \end{bmatrix},$$

and

$$m_I = |x|\sqrt{(R - |x|)^2 + y^2}.$$

Here, the initial guess for the $\mathbf{Q}$-tensor is computed from a simplified two-dimensional mathematical representation of a family of tori, and we have taken the major radius $R = 0.5$ in this implementation.

Furthermore, we specify the values of parameters in this experiment:

$$a_1 = -10, \ a_2 = 0, \ a_3 = 10, \ B = 10^{-5}, \ K = 0.3,$$
$$q = 30, \text{ and } l = 30.$$

The total energy to be minimised in this scenario is

$$\mathcal{J}_\epsilon(u, \mathbf{Q}) = \int_\Omega \left(\frac{a_1}{2} (u)^2 + \frac{a_2}{3} (u)^3 + \frac{a_3}{4} (u)^4 \right.$$

$$+ B \left| \mathcal{D}^2 u + q^2 \left(\mathbf{Q} + \frac{\mathbf{I}_2}{2} \right) u \right|^2$$

$$+ \frac{K}{2} |\nabla \mathbf{Q}|^2 - l \left(\mathrm{tr} \left(\mathbf{Q}^2 \right) \right) + l \left(\mathrm{tr} \left(\mathbf{Q}^2 \right) \right)^2 \right)$$

$$+ \sum_{e \in \mathcal{E}_I} \int_e \frac{1}{2h_e^3} ([\![\nabla u]\!])^2. \tag{10.2}$$

We present the bifurcation diagram in Figure 10.1 for this scenario and quantitatively determine which of these solutions is the ground state as a function of θ_0. To give more details on those solution branches with the lowest energy in the bifurcation diagram, we show some computed stationary states in Figure 10.2 as a function of θ_0 by minimising (10.2). For each state, we display the value of the energy functional

$$\mathcal{J}(u, \mathbf{Q}) = \int_\Omega \left(\frac{a_1}{2} (u)^2 + \frac{a_2}{3} (u)^3 + \frac{a_3}{4} (u)^4 \right.$$

$$+ B \left| \mathcal{D}^2 u + q^2 \left(\mathbf{Q} + \frac{\mathbf{I}_2}{2} \right) u \right|^2$$

$$+ \frac{K}{2} |\nabla \mathbf{Q}|^2 - l \left(\mathrm{tr} \left(\mathbf{Q}^2 \right) \right) + l \left(\mathrm{tr} \left(\mathbf{Q}^2 \right) \right)^2 \right),$$

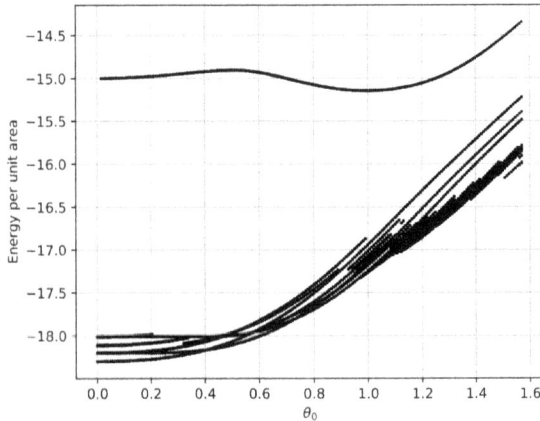

Fig. 10.1. The bifurcation diagram of the defect-free scenario.

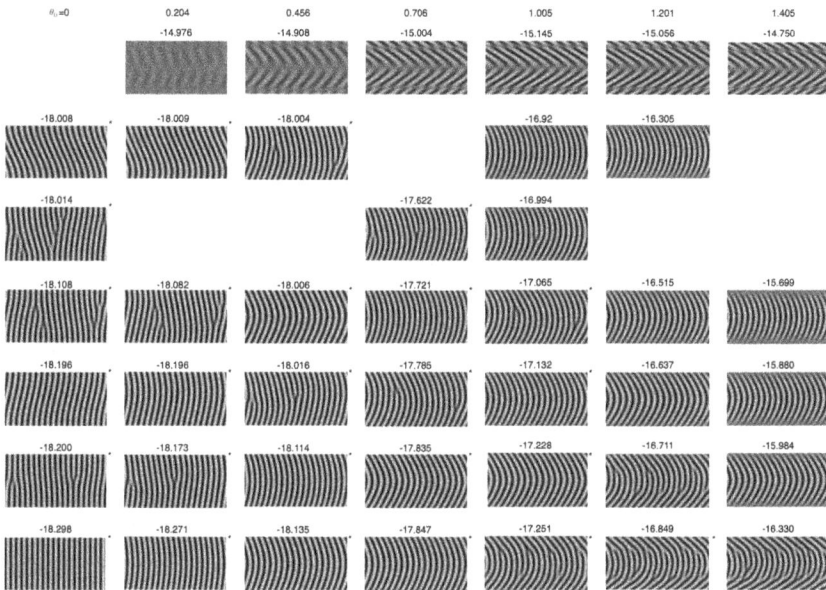

Fig. 10.2. Stationary states obtained at different values of θ_0 in the defect-free scenario. The visualisation displays the density perturbation u. For each solution, the value of the energy functional per unit area is displayed above it and we specify the stable profiles with asterisks. The bottom row depicts the lowest energy solution found for each value of θ_0.

per unit area. For each column (i.e., fixed value of θ_0), we organise the stationary states in an energy-decreasing order and identify stable profiles with asterisks. The bottom row depicts the lowest-energy minimisers found, all of which are stable.

We can observe from Figure 10.2 an energetic competition between the cost of bending and the cost of introducing disclinations from those equilibrium structure as a function of θ_0. More specifically, when $\theta_0 = 0$ (thus the boundary conditions enforce that the director $\mathbf{n}_e$ is horizontally aligned), the resulting configuration is with the layers extending vertically between the substrates in the "bookshelf" geometry. As θ_0 is increased from zero, the boundary conditions impose a bend deformation on the smectic. This can be accommodated in several ways: by distributing the deformation over the vertical direction (see the second picture in the bottom row of Figure 10.2); by localising the bend to a region in the center with the layers flat and tilted in opposite directions in the top and bottom of the domain (see the third and fourth pictures in the bottom row of Figure 10.2); or by introducing edge disclinations to relieve the cost of elastic deformation (see the last three pictures in the bottom row of Figure 10.2).

We also include one video *scenario-i-lowest-energy-in-theta-zero.mp4* in Xia (2021c) to illustrate the stationary configurations of lowest energy found as we vary the applied bend deformation $\theta_0 \in [0, \pi/2]$. The profiles shown in this video are all stable.

10.3   Scenario II: Focal Conic Domains

Among all defect structures in smectic liquid crystals, the most common one is focal conic domains (FCDs, as illustrated in Williams and Kléman (1975, Fig. 1)): the smectic layers are kept equidistant and parallel, with common normals and same center of curvature along the same normal. Such smectic layers are examples of *Dupin cyclides* which present two types of disclinations: ellipses and hyperbolas (also known as the *fonal conics*). When the ellipse degenerates to a circle and the hyperbola to a straight line, these smectic layers are called toroidal focal conic domains (TFCDs). In this section, we simulate FCDs and TFCDs using our proposed model (8.10).

We discretise the cuboid $\Omega = [-1.5, 1.5] \times [-1.5, 1.5] \times [0, 2]$ into $6 \times 6 \times 5$ uniform hexahedra, to avoid a directional bias observed in numerical solutions with tetrahedra. To simulate TFCDs or FCDs, we must impose boundary conditions (weakly or strongly) that respect their physical properties. To this end, we label the six boundary faces of Ω as

$$\Gamma_{\text{left}} = \{(x, y, z) : x = -1.5\}, \quad \Gamma_{\text{right}} = \{(x, y, z) : x = 1.5\},$$

$$\Gamma_{\text{back}} = \{(x, y, z) : y = -1.5\}, \quad \Gamma_{\text{front}} = \{(x, y, z) : y = 1.5\},$$

$$\Gamma_{\text{bottom}} = \{(x, y, z) : z = 0\}, \quad \Gamma_{\text{top}} = \{(x, y, z) : z = 2\},$$

and consider the following surface energy:

$$F_{\text{surface}}(\mathbf{Q}) = \int_{\Gamma_{\text{bottom}}} \frac{w}{2} |\mathbf{Q} - \mathbf{Q}_{\text{radial}}|^2 + \int_{\Gamma_{\text{top}}} \frac{w}{2} |\mathbf{Q} - \mathbf{Q}_{\text{vertical}}|^2, \quad (10.3)$$

where w denotes the weak anchoring weight,

$$\mathbf{Q}_{\text{radial}} = \begin{bmatrix} \frac{x^2}{x^2+y^2} - \frac{1}{3} & \frac{xy}{x^2+y^2} & 0 \\ \frac{xy}{x^2+y^2} & \frac{y^2}{x^2+y^2} - \frac{1}{3} & 0 \\ 0 & 0 & -\frac{1}{3} \end{bmatrix}$$

represents an in-plane (x–y plane) radial configuration of the director, and

$$\mathbf{Q}_{\text{vertical}} = \begin{bmatrix} -\frac{1}{3} & 0 & 0 \\ 0 & -\frac{1}{3} & 0 \\ 0 & 0 & \frac{2}{3} \end{bmatrix}$$

gives a vertical (i.e., along the z-axis) alignment configuration of the director. Therefore, the final form of the functional to be minimised in the TFCD scenario is

$$
\begin{aligned}
\mathcal{J}_\epsilon(u, \mathbf{Q}) = \int_\Omega & \left(\frac{a}{2}(u)^2 + \frac{b}{3}(u)^3 + \frac{c}{4}(u)^4 \right. \\
& + B \left| \mathcal{D}^2 u + q^2 \left(\mathbf{Q} + \frac{\mathbf{I}_3}{3} \right) u \right|^2 + \frac{K}{2} |\nabla \mathbf{Q}|^2 \\
& \left. - \frac{l}{2}(\operatorname{tr}(\mathbf{Q}^2)) - \frac{l}{3}(\operatorname{tr}(\mathbf{Q}^3)) + \frac{l}{2}(\operatorname{tr}(\mathbf{Q}^2))^2 \right) \\
& + \int_{\Gamma_{\text{bottom}}} \frac{w}{2} |\mathbf{Q} - \mathbf{Q}_{\text{radial}}|^2 + \int_{\Gamma_{\text{top}}} \frac{w}{2} |\mathbf{Q} - \mathbf{Q}_{\text{vertical}}|^2 \\
& + \sum_{e \in \mathcal{E}_I} \int_e \frac{1}{2 h_e^3} (\llbracket \nabla u \rrbracket)^2.
\end{aligned}
\tag{10.4}
$$

For the FCD scenario, we only change the top boundary condition to perturb the preferred tilted director configuration. We perturb the angle θ_c between the director and the z-axis on the top surface Γ_{top}, thus adopting

$$\mathbf{Q}_c = \begin{bmatrix} -\frac{1}{3} & 0 & 0 \\ 0 & (\sin(\theta_c))^2 - \frac{1}{3} & \sin(\theta_c)\cos(\theta_c) \\ 0 & \sin(\theta_c)\cos(\theta_c) & (\cos(\theta_c))^2 - \frac{1}{3} \end{bmatrix}$$

instead of $\mathbf{Q}_{\text{vertical}}$ in (10.4). Note that when taking $\theta_c = 0$, we return to the TFCD case.

Furthermore, we take the initial guesses:

$$u = \cos(6\pi z), \quad \mathbf{Q} = \mathbf{Q}_{ic},$$

where $\mathbf{Q}_{ic} = \left(\mathbf{n}_{ic} \otimes \mathbf{n}_{ic} - \frac{\mathbf{I}_3}{3} \right)$ with

$$\mathbf{n}_{ic} = \frac{1}{m_{ic}} \begin{bmatrix} x \left(\sqrt{x^2 + y^2} - R \right) \\ y \left(\sqrt{x^2 + y^2} - R \right) \\ z \left(\sqrt{x^2 + y^2} \right) \end{bmatrix},$$

and

$$m_{ic} = \sqrt{x^2 + y^2}\sqrt{\left(R - \sqrt{x^2 + y^2}\right)^2 + z^2}.$$

Here, the initial guess for the **Q**-tensor is computed from the mathematical representation for a family of tori, and we have taken a major radius $R = 1.5$ in our implementation.

We specify the values of parameters used in the (T)FCD experiments:

$$a_1 = -10, \ a_2 = 0, \ a_3 = 10, \ R - 10^{-3}, \ K - 0.03,$$
$$q = 10, \ l = 30 \text{ and } w = 10.$$

Two numerical solutions of simulating TFCDs are given in Figure 10.3. One can see that these zero isosurfaces of density indeed present a physically reasonable TFCD with two parts of singularities: circles at the bottom and the central line along the cusps. Notice that we are not imposing any periodic conditions of the density u but only weakly enforcing boundary conditions as in (10.3) on the tensor field **Q**. It turns out in Figure 10.3 that the smectic layers align themselves to the director field arising from **Q** and thus the periodicity on the lateral faces can be observed. This is due to the coupling term in the model. Other than the TFCD solution as illustrated in Figure 10.3, it also shows another possibility of equilibrium solution with single screw dislocation at the central line, though a theoretical investigation of such interesting structure remains an open problem. We further comment that the single screw dislocation possesses higher energy value than that of the TFCD solution. At this point we are not sure if such

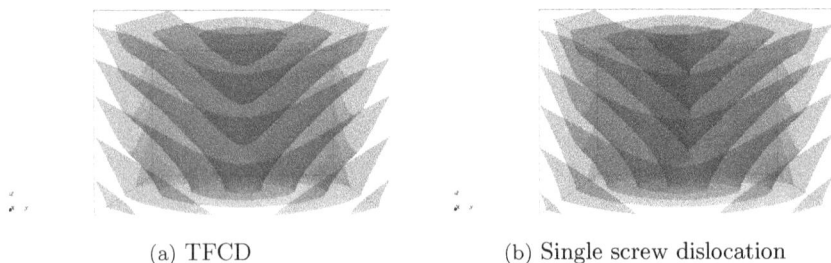

(a) TFCD   (b) Single screw dislocation

Fig. 10.3. (a) The first converged solution using Newton's method on a mesh of $6 \times 6 \times 5$ hexahedra using the TFCD settings. (b) Another solution profile with single screw dislocation around the central axis of the cuboid. The solution with screw dislocation has higher energy and both are stable. The gray layers are zero iso-surfaces of the density variation u.

a dislocation is physically realistic, but it presents an interesting pattern of defects in this numerical experiment.

In addition, we noticed from some preliminary experiments under the TFCD problem settings that a special case, i.e., the radial configuration of director molecules, of the planar anchoring condition is more likely to give a successful presentation of TFCDs. This may be helpful for a better and more accurate understanding of realistic boundary conditions to be enforced for the appearance of TFCDs.

The TFCD profile shown in Figure 10.3 can be generalised into an asymmetric version, thus presenting the Dupin cyclides. We take $\theta_c = \frac{\pi}{12}$ and run the experiment with the other parameters chosen as in the TFCD settings. Three solution examples are shown in Figure 10.4, which includes an FCD solution Figure 10.4(a), a single screw dislocation Figure 10.4(b) and a double screw dislocation structure Figure 10.4(c). They are all stable solutions. It can be observed in the FCD solution profile that the smectic layers have deformed asymmetrically when responding to the tilting of the director on the top face. Note here that the FCD solution has the lowest energy due to the energy cost of the dislocation defects. To depict these three solution structures more closely, we further present an additional

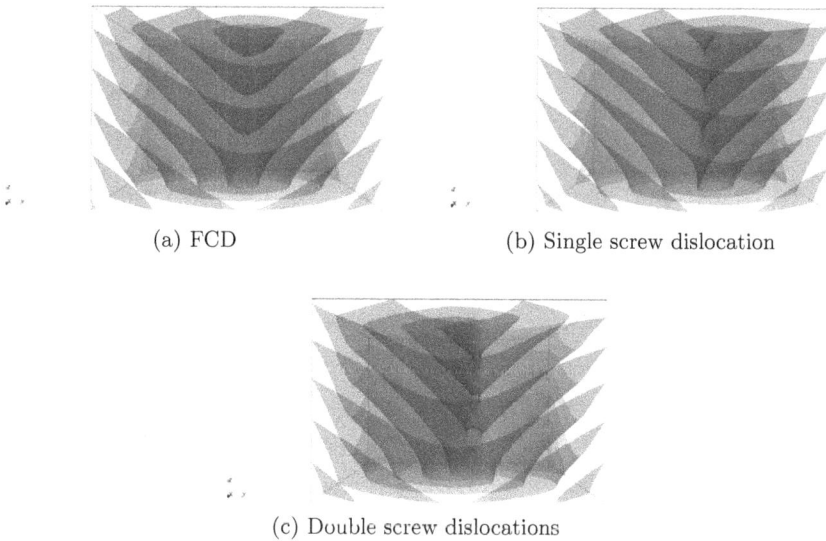

(a) FCD   (b) Single screw dislocation

(c) Double screw dislocations

Fig. 10.4. Three numerical solutions for $\theta_c = \frac{\pi}{12}$ on a mesh of $6 \times 6 \times 5$ hexahedra. The solution with double screw dislocations has highest energy while the FCD solution possesses lowest energy. All profiles are stable.

video *scenario-ii-pi12.mp4* in Xia (2021c), describing the zero-isosurfaces of the smectic density variation field u and colouring the isosurfaces by height (the z-coordinate) to assist with depth perception. The time axis of the video is used to illustrate the internal structure of the layers.

If we take $\theta_c = \frac{\pi}{10}$, the first converged solution shows a FCD structure as presented in Figure 10.5(a). Another example is also given in Figure 10.5(b) which yields a single screw dislocation profile possessing higher energy. Again, both profiles are stable equilibrium points of the energy (10.4).

Moreover, as we increase the value of θ_c to be $\frac{\pi}{8}$, two examples of stable numerical solutions are shown in Figure 10.6, where the focal conic curve in the FCD solution tilts more when compared with that in Figure 10.4(a). We also see the screw dislocation structure possessing higher energy than that of the FCD solution in this experiment.

As the Dupin cyclide has a confocal pair of a hyperbola and an ellipse, we fit a hyperbola to each solution with least squares (data points extracted via ParaView (Ahrens *et al.*, 2005)) and calculate its eccentricity (e.g., for a hyperbola expressed as $\frac{y^2}{a_{fit}^2} - \frac{z^2}{b_{fit}^2} = 1$, its eccentricity is defined

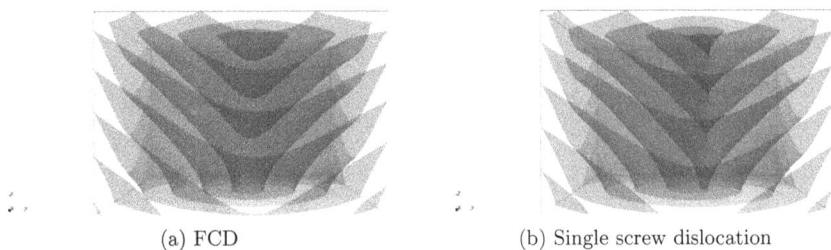

| (a) FCD | (b) Single screw dislocation |

Fig. 10.5. Two solution profiles by taking $\theta_c = \frac{\pi}{10}$ on a mesh of $6 \times 6 \times 5$ hexahedra. The solution with screw dislocation has higher energy. Both profiles are stable.

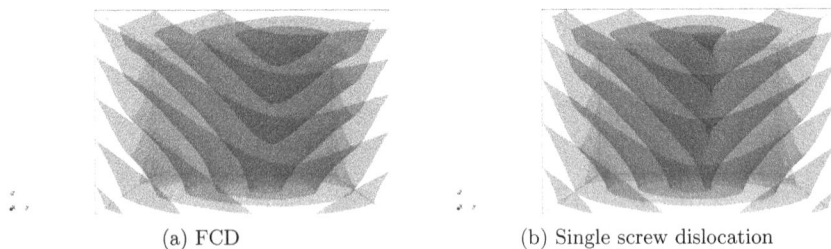

| (a) FCD | (b) Single screw dislocation |

Fig. 10.6. Two numerical solutions for $\theta_c = \frac{\pi}{8}$ on a mesh of $6 \times 6 \times 5$ hexahedra. The solution with screw dislocation has higher energy. Both profiles are stable.

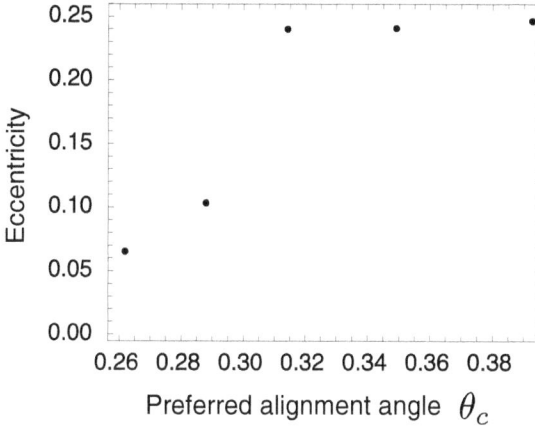

Fig. 10.7. Eccentricity of FCD solutions as a function of preferred surface alignment angle.

as $\frac{\sqrt{a_{fit}^2+b_{fit}^2}}{a_{fit}}$). Then the eccentricity of the ellipse is the inverse of that of the confocal hyperbola. Values of eccentricity fitted from the solution set are shown as a function of the preferred surface alignment angle θ_c in Figure 10.7.

10.4 Scenario III: Oily Streaks

Besides the (T)FCD defects illustrated in the previous section, there is another type of defects that are experimentally observable in films of 8CB deposited in air on crystalline surfaces of molybdenite (MoS_2) (Michel *et al.*, 2004): the so-called *oily streaks* (OS). When thin smectic liquid crystal films are subject to competing boundary conditions, they can form interesting patterns. In particular, *planar degenerate anchoring* (i.e., the molecules on the surface are in the plane of the surface) and *homeotropic anchoring* (i.e., the molecules prefer to be perpendicular to the surface) imposed on two opposing surfaces can form a periodic stacking of flattened hemi-cylinders, as shown in Michel *et al.* (2004, Figs. 9 and 16). We simulate this typical defect in this section using our proposed model (8.10).

Let r denote the aspect ratio of a rectangle $\Omega = [-r, r] \times [0, 2]$ with the boundaries labels

$$\Gamma_l = \{(x, y) : x = -r\}, \quad \Gamma_r = \{(x, y) : x = r\},$$
$$\Gamma_b = \{(x, y) : y = 0\}, \quad \Gamma_t = \{(x, y) : y = 2\}.$$

We impose the following surface energy:

$$F_{\text{surface}}(\mathbf{Q}) = \int_{\Gamma_b} \frac{w}{2} |\mathbf{Q} - \mathbf{Q}_{\text{bottom}}|^2 + \int_{\Gamma_t \cup \Gamma_l \cup \Gamma_r} \frac{w}{2} |\mathbf{Q} - \mathbf{Q}_{\text{top}}|^2,$$

where w is the weak anchoring weight and two weakly prescribed configurations $\mathbf{Q}_{\text{bottom}}$ and $\mathbf{Q}_{\text{top}}$ are given by

$$\mathbf{Q}_{\text{bottom}} = \begin{bmatrix} \frac{1}{2} & 0 \\ 0 & -\frac{1}{2} \end{bmatrix},$$

yielding horizontally aligned directors, and

$$\mathbf{Q}_{\text{top}} = \begin{bmatrix} -\frac{1}{2} & 0 \\ 0 & \frac{1}{2} \end{bmatrix},$$

yielding vertically aligned directors.

In this experiment, we always discretise the domain Ω into 90×30 quadrilateral elements, even as we change the domain size by varying the aspect ratio r. The final form of the functional to be minimised in this scenario is

$$\begin{aligned}
\mathcal{J}_\epsilon(u, \mathbf{Q}) = \int_\Omega \Big(&\frac{a_1}{2}(u)^2 + \frac{a_2}{3}(u)^3 + \frac{a_3}{4}(u)^4 \\
&+ B \left| \mathcal{D}^2 u + q^2 \left(\mathbf{Q} + \frac{\mathbf{I}_2}{2} \right) u \right|^2 \\
&+ \frac{K}{2}|\nabla \mathbf{Q}|^2 - l\left(\mathrm{tr}\left(\mathbf{Q}^2\right)\right) + l\left(\mathrm{tr}\left(\mathbf{Q}^2\right)\right)^2 \Big) \\
&+ \int_{\Gamma_b} \frac{w}{2} |\mathbf{Q} - \mathbf{Q}_{bottom}|^2 + \int_{\Gamma_t \cup \Gamma_l \cup \Gamma_r} \frac{w}{2} |\mathbf{Q} - \mathbf{Q}_{top}|^2 \\
&+ \sum_{e \in \mathcal{E}_I} \int_e \frac{1}{2h_e^3} (\llbracket \nabla u \rrbracket)^2.
\end{aligned} \tag{10.5}$$

We take the same form of the initial guesses for u and $\mathbf{Q}$ as in (10.1) but with a larger major radius $R = 1$ in this scenario.

Finally, we specify the values of parameters in this experiment:

$$a_1 = -10, \ a_2 = 0, \ a_3 = 10, \ B = 10^{-5}, \ K = 0.3,$$
$$q = 30, \ l = 1 \text{ and } w = 10.$$

Based on X-ray diffraction experiments of thin smectic films, Michel *et al.* (Lacaze *et al.*, 2007) proposed some approximate structures of oily

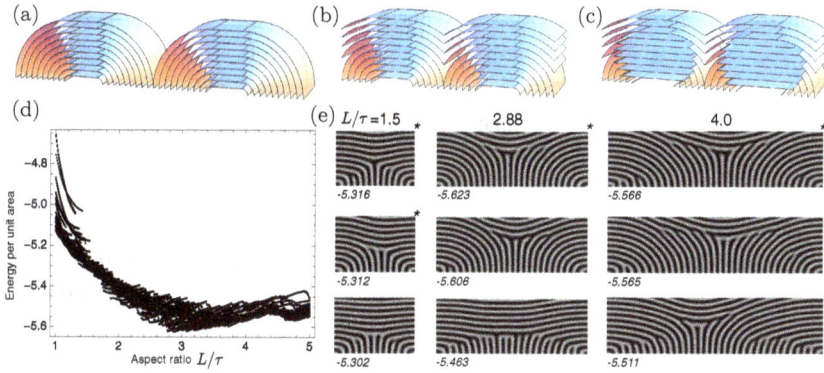

Fig. 10.8. Oily streaks. (a)–(c) Candidate structures proposed in Michel *et al.* [Lacaze *et al.* (2007)] consistent with X-ray diffraction. (d) Bifurcation diagram of structures as a function of aspect ratio. (e) Selected stationary states obtained at different aspect ratio *r*. The top row represents the lowest energy solution found. For each solution, the value of the energy functional per unit area is displayed below it with asterisks indicating stable profiles.

streaks as illustrated in Figures 10.8(a)–10.8(c). Since some experiments reveal that the smectic layer normals are continuously oriented for smectic layers that are parallel to the plane of substrate for thin films, the authors gave a possible structure in Figure 10.8(a) depicting periodic units incorporating sections of cylinders joined to planes oriented parallel to the substrate. However, this structure implies significant deformations of the free interface with singular points between units. To avoid so, they proposed a more complex structure as illustrated in Figure 10.8(b) incorporating curvature walls between units. Moreover, it is observed in the X-ray diffraction of even thinner films that an apparent excess of the planar region is shown, which cannot be explained by either structure discussed so far Michel *et al.* (2006). Therefore, Figure 10.8(c) provides a possible structure consistent with the experimental data envisioned in Michel *et al.* (2006), though it is energetically very costly.

By implementing the proposed mathematical model, we display the partially enumerated energy landscape in Figure 10.8(d), showing an extremely dense thicket of solutions. This qualitatively supports earlier work in that an overall minimiser occurs at an aspect ratio of around 3, which is similar to experimental values even with no parameter tuning performed here. Close examination of the energy landscape, together with the corresponding solution set, shows many small discontinuous jumps that result from delicate commensurability effects, whereby certain sizes of domain are compatible with a given periodicity of the layers as well as from variations in the number of defects and their detailed placement. Similar effects have been observed when other periodic liquid crystals such as cholesterics are confined in domains that promote geometric frustration (Emerson *et al.*, 2018).

The solution set obtained contains examples reminiscent of previously proposed structures (Figure 10.8(e)). The minimum energy states found at different aspect ratio contain cylindrical sections mediated by a defect-filled region reminiscent of the mesoscopic rotating grain boundaries. Other solutions displayed in the lowest row of Figure 10.8(e) are quite different from those heretofore proposed, where regions of relatively vertically oriented layers sit atop cylindrical regions interspersed with defects. Each of these incorporates a greater proportion of vertical layers relative to the hemicylindrical-planar *ansatz* of Figures 10.8(a) and 10.8(b) and may provide alternative structures for oily streaks in ultrathin films. In future work, the boundary conditions at the top interface should be carefully reconsidered, including the incorporation of a free interface.

We refer readers to the video *scenario-iii-lowest-energy-in-r.mp4* (Xia, 2021c) depicting the lowest-energy configurations discovered as we vary the aspect ratio $r \in [1, 5]$. All presented profiles in this video are stable.

10.5   Summary

In this chapter, we simulated three smectic scenarios involving boundary conditions that are incompatible with uniform smectic order to investigate the effectiveness of our proposed mathematical model (8.10) in characterising the defect structures, e.g., (toroidal) focal conic domains and oily streaks in smectics. Our new model successfully reproduced, even without careful tuning of parameters, a number of experimentally observed and theoretically expected phenomena, as well as producing new candidate structures for thin smectic films that are explicitly stationary states of an energy functional. We believe this success can lead to many other smectic applications in future.

Part 4

Conclusions and Perspectives

Chapter 11

Summary and Conclusions

11.1 Conclusions

This book tackles and implements several energy minimisation problems arising from modelling cholesteric liquid crystals, ferronematics and smectic-A liquid crystals.

In Chapters 2–4, we consider the Oseen–Frank model of nematic and cholesteric liquid crystals that employs a vector-valued director field as state variable, subject to a unit-length constraint. We apply augmented Lagrangian methods to transform the constrained minimisation problem into an unconstrained one of saddle point type. The benefits of the AL method are twofold: it helps control the Schur complement, enabling fast solvers; and it improves the discrete constraint as we increase the value of the penalty parameter in the implementation. The details of the relevant discussions are illustrated in Chapter 2. The tradeoff is that it complicates the solution of the top-left director block, as it adds a semi-definite term with a large coefficient arising from the AL formulation. To resolve this issue, our core contribution in Chapter 3 is to develop a robust and efficient multigrid solver. A parameter-robust relaxation method is achieved by developing a space decomposition that stably captures the kernel of the semi-definite terms. Chapter 4 demonstrates the validity of our derived parameter- and mesh-independent solver through several numerical experiments.

Due to the difficulties of (i) solving a constrained minimisation problem and (ii) representing certain defect structures (e.g., half charge defects), we turn from the Oseen–Frank theory to the Landau–de Gennes modelling theory that uses a tensor-valued state variable. We consider a one-dimensional model of ferronematics in Chapters 5 and 7 to study order reconstruction solutions, bifurcations, and multistability. Chapter 6 illustrates some simple examples using the deflation technique to compute multiple solutions. We then construct a novel numerical bifurcation analysis in Chapter 7 of theoretical results analysed in Dalby *et al.* (2022) and perform an asymptotic analysis (see Section 7.4) for certain model parameters. We pay special attention to defect structures (domain walls in ferronematics) in our investigation. These numerical studies form a solid basis for validating analytical results and demonstrate the promising potential of capturing defects using the Q-tensor theory.

In the third part of this book (Chapters 8–10), we devote ourselves to proposing a new continuum mathematical model for smectic-A liquid crystals, and developing a convergent finite element discretisation thereof. To represent half charge defects that are likely to happen in smectics, the model is characterised by a tensor-valued nematic order parameter and a real-valued smectic order parameter. We prove an existence result in Chapter 8 for the proposed minimisation problem. Chapter 9 investigates an appropriate finite element formulation for solving the optimality conditions, which are essentially a coupled system involving a fourth-order PDE and a second-order PDE. For the fourth-order problem, we take the common Lagrange elements with an interior penalisation term to avoid the use of more complicated H^2-conforming elements. The second-order PDE, which comes from the classical Landau–de Gennes model for nematic phases, is simpler and is discretised with standard Lagrange elements. This chapter derives some *a priori* error estimates for both variables in the decoupled case, accompanied by numerical verifications of convergence rates in the coupled case. Some interesting applications of the new model are presented in Chapter 10, where some typical defect structures are numerically captured for the first time. This shows promise for further related work in smectic liquid crystals.

11.2   Potential Future Work

11.2.1   *Future work I*

Regarding the Oseen–Frank model, we have developed in Chapter 3 the theory for the construction of a robust multigrid algorithm for the equal-constant nematic LC. Extensions to the multi-constant case give rise to some additional difficulties, especially in the characterisation of the kernels of the $\nabla\cdot$ and $\nabla\times$ operators in the Frank energy density (2.3). A potential resolution for this difficulty is to use the de Rham complexes (Arnold *et al.*, 2000). The smooth de Rham complex in two dimensions is given by

$$\mathbb{R} \xrightarrow{\text{id}} C^\infty(\Omega) \xrightarrow{\nabla\times} [C^\infty(\Omega)]^2 \xrightarrow{\nabla\cdot} C^\infty(\Omega) \xrightarrow{\text{null}} 0,$$

where the kernel Ker($\cdot$) of an operator is the range Range($\cdot$) of the preceding operator on a simply connected domain. For instance, Range($\nabla\times$) = Ker($\nabla\cdot$). This allows us to characterise the divergence-free vector fields as the curls of potentials. However, the above de Rham complex is rather restrictive in implementation as it requires smooth spaces. For our interests in LC problems with directors having H^1-regularity, we should instead utilise complexes involving Sobolev spaces, e.g., the so-called Stokes complex in two dimensions:

$$\mathbb{R} \xrightarrow{\text{id}} H^2(\Omega) \xrightarrow{\nabla\times} [H^1(\Omega)]^2 \xrightarrow{\nabla\cdot} L^2(\Omega) \xrightarrow{\text{null}} 0.$$

Discrete versions of these complexes are much harder to construct and often result in high order polynomials due to the high regularity requirements, such as the H^2-regularity. The study of an appropriate de Rham complex will help characterise the kernel of $\nabla\cdot$ and $\nabla\times$ operators in the finite element spaces. This will allow for the preconditioner developed in this book to be analysed for the multi-constant case.

11.2.2 *Future work II*

With the success in predicting typical defects in smectic-A liquid crystals, we can extend our result to encompass the smectic-C phase, and thus give a unified model for liquid crystals including isotropic, nematic, smectic-A and smectic-C phase transitions.

The idea can be built on the work of Biscari *et al.* (2007), who present a de Gennes variational theory based on a complex-valued smectic order parameter ψ and a tensor-valued nematic order parameter $\mathbf{Q}$ to simultaneously describe those transitions. More specifically, the difference between smectic-A and C phases is characterised by a new interaction term

$$\chi := \mathbf{Q}\nabla\psi \times \nabla\psi. \tag{11.1}$$

If the nematic director is aligned to the smectic layer normals as in the smectic-A phase, then $\chi = 0$, otherwise a nonzero χ represents a smectic-C phase. The following energy from the interaction term characterising smectic-C phases is added to the free energy:

$$\int_\Omega e_{AC}\boldsymbol{\chi}\cdot\boldsymbol{\chi} = \int_\Omega e_{AC}\left|\mathbf{Q}\nabla\psi\times\nabla\psi\right|^2, \tag{11.2}$$

where e_{AC} is a constant. Note that a negative value of e_{AC} will enforce smectic-C phases in the model and a positive value results in smectic-A phases.

Considering our proposed model of smectic-A LC in Chapter 8, which is based on a real-valued smectic density u and a tensor-valued nematic order parameter $\mathbf{Q}$, we intend to introduce the following interaction term similar to (11.1) to distinguish the smectic-A and C phases:

$$\boldsymbol{\chi} = \mathbf{Q}\nabla u\times\nabla u,$$

and add

$$\int_\Omega \frac{e_{AC}}{2}\left|\mathbf{Q}\nabla u\times\nabla u\right|^2$$

to our proposed free energy (8.10).

One important potential application could be simulating smectic-C LC in a wedge, as illustrated in Carlsson *et al.* (1991, Section 3), where smectic layers are expected to form concentric cylinders with the common axis coinciding with the center of the wedge. This simulation is used there to examine different distortion effects existed in smectic-C LC. Another avenue to pursue is to investigate the chevron structure (see Biscari *et al.*, 2007, Section IV), one of the most interesting defects existing in the smectic-C phase.

11.2.3 *Future work III*

Concerning the smectic-A phase, there are several topics that can be pursued further using our proposed smectic model (8.10).

The computational time required to solve three-dimensional problems is noticeable longer than for two-dimensional problems. This motivates the use of a faster algorithm to improve computational efficiency. Some choices can be taken, e.g., designing a preconditioner for the model (8.10) or using the static condensation technique (Guyan, 1965; Irons, 1965) to reduce the size of the stiffness matrix. Moreover, due to the similarities of our adopted $\mathcal{C}^0$-IP methods and the weakly over-penalised symmetric interior penalty method illustrated in Brenner *et al.* (2010) for biharmonic problems, we may build on Brenner *et al.* (2010) for the construction and analysis of efficient solvers for the smectic-smectic block of the matrix.

Since our proposed model characterises both nematic and smectic-A phases, it may be used to investigate the nematic-smectic transition by varying the temperature-dependent parameter a_1. Zappone *et al.* recently confirmed the existence of intermediate LC state analogous to superconductors (Zappone *et al.*, 2020) for thin smectic films of different thicknesses. In particular, they find the so-called P-texture only observed when cooling a thin smectic film. It can be seen from this schematic description that the $-\frac{1}{2}$ defects possess similar structures of defect walls as in the oily streaks problem explored in Section 10.4. This motivates us to apply our new model to study the nematic–smectic transition.

From the numerical perspective and inspired by the progress of using our proposed smectic-A model (8.10) to capture typical defects in smectics, we believe it can be further applied to more laboratory experiments to help in investigating internal defect structures. For instance, one could use our smectic model to characterise and analyse edge and screw dislocations in a wedge similarly to Lelidis *et al.* (2006). We give a preliminary result (see Figure 11.1) related to this wedge problem as described in Lelidis *et al.* (2006, Figure 2).

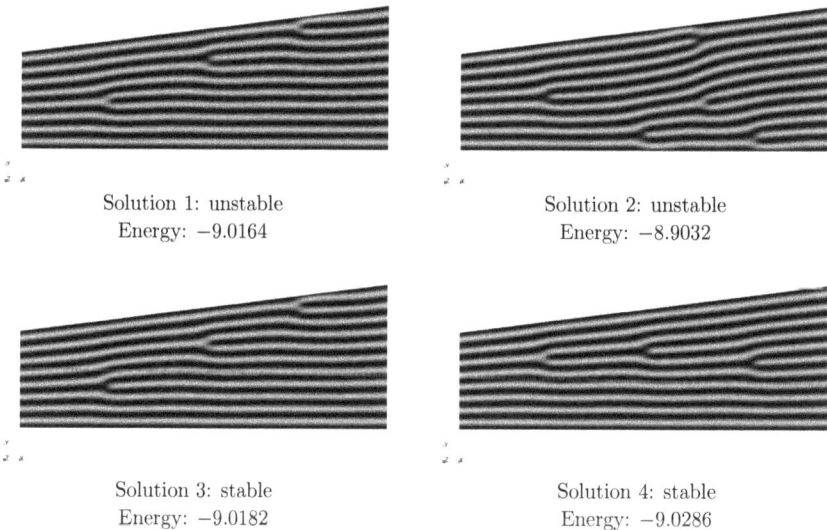

Solution 1: unstable
Energy: -9.0164

Solution 2: unstable
Energy: -8.9032

Solution 3: stable
Energy: -9.0182

Solution 4: stable
Energy: -9.0286

Fig. 11.1. Four solution profiles and their stabilities with strongly-enforced Dirichlet data on $\delta\rho = 1$ and strongly-enforced homeotropic boundary conditions of $\mathbf{Q}$ on top and bottom surfaces of a wedge. Solution 4 with three edge dislocations has the lowest energy.

Another avenue of investigation is to compare results from our model with actual experiments and with simulations conducted using other methods (particularly Monte Carlo and Density Functional Theory). This would yield a better understanding of the strengths and weaknesses of the different available smectic modelling theories. We have begun to collaborate with the authors of Wittmann *et al.* (2021) to investigate the smectic structures that are predicted by different modelling frameworks in confined geometries with holes. A simple example of a geometry to be considered in this work is two overlapped annuli, as illustrated in Figure 11.2. We present some preliminary results (see Figures 11.3 and 11.4) of obtained profiles

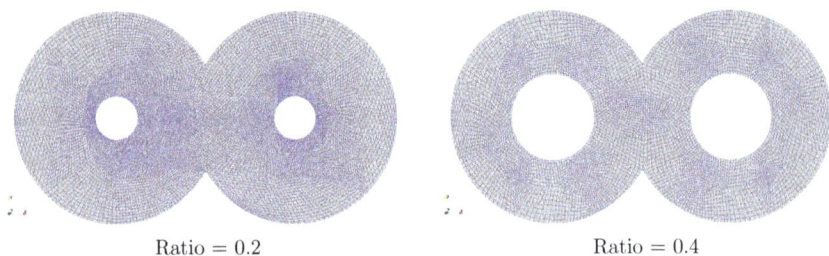

Ratio = 0.2   Ratio = 0.4

Fig. 11.2. Meshes of two fused annuli. The domains differ in the sizes of the inclusions.

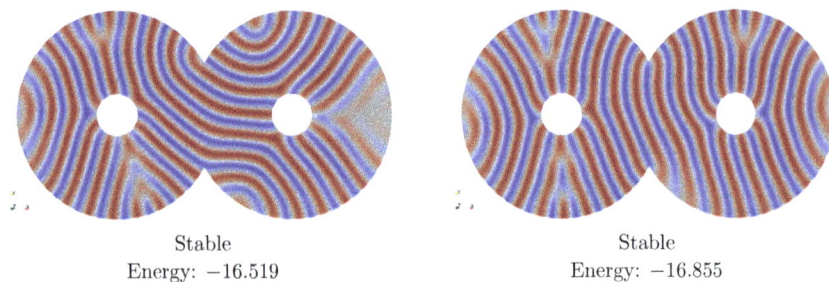

Stable   Stable
Energy: −16.519   Energy: −16.855

Fig. 11.3. Two solution profiles of the geometry with inclusion ratio 0.2.

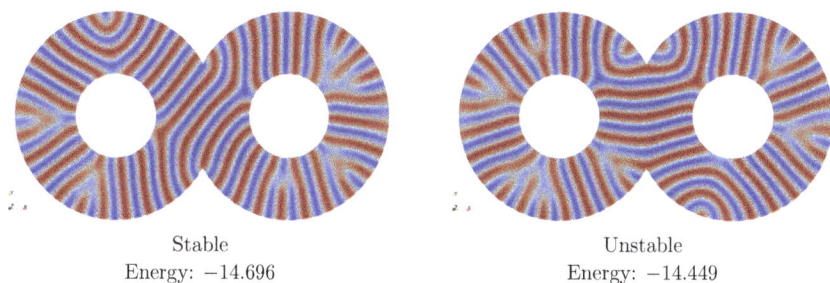

Stable
Energy: −14.696

Unstable
Energy: −14.449

Fig. 11.4. Two solution profiles for the geometry with inclusion ratio 0.4.

when tangential boundary conditions are imposed along both external and inner circles of the annuli. As of writing, laboratory experiments in these geometries are underway, led by Prof. Dirk Aarts of the Oxford Colloid Group.

Appendix A

A.1 Stability of the Newton System

Recall that the rth Newton iteration (6.22) is

$$Da(\mathbf{u}^r, \mathbf{v})[\delta\mathbf{u}] - \int_{\Gamma_D} D\mathbf{t}(\mathbf{u})[\delta\mathbf{u}] \cdot \mathbf{v} \, ds + \gamma \int_{\Gamma_D} \delta\mathbf{u} \cdot \mathbf{v} \, ds = -R^*(\mathbf{u}^r, \mathbf{v}),$$

$$(A.1)$$

with

$$Da(\mathbf{u}^r, \mathbf{v})[\delta\mathbf{u}] := \int_{\Omega} \nabla\delta\mathbf{u} : \mathbf{C}(\mathbf{u}^r) : \delta\mathbf{v} \, dx.$$

We denote the left-hand side bilinear form by $A(\mathbf{u}^r; \mathbf{v}, \delta\mathbf{u})$ and the L^2-norm over the Dirichlet boundary Γ_D by $\|\cdot\|_{0,\Gamma_D}$. To ensure the solvability of the above Newton system, the major issue is to prove that with the current approximation $\mathbf{u}^r \in V$,

$$A(\mathbf{u}^r; \mathbf{w}, \mathbf{w}) > 0$$

for all nonzero $\mathbf{w} \in V$. In the following, we will ignore the superscript r for notational simplicity.

Note that

$$
\begin{aligned}
A(\mathbf{u}; \mathbf{w}, \mathbf{w}) &= Da(\mathbf{u}, \mathbf{w})[\mathbf{w}] - \int_{\Gamma_D} D\mathbf{t}(\mathbf{u})[\mathbf{w}] \cdot \mathbf{w} \, ds + \gamma \int_{\Gamma_D} \mathbf{w} \cdot \mathbf{w} \, ds \\
&\geq Da(\mathbf{u}, \mathbf{w})[\mathbf{w}] - \|D\mathbf{t}(\mathbf{u})[\mathbf{w}]\|_{0,\Gamma_D} \|\mathbf{w}\|_{0,\Gamma_D} + \gamma\|\mathbf{w}\|_{0,\Gamma_D}^2 \\
&\geq Da(\mathbf{u}, \mathbf{w})[\mathbf{w}] - C\sqrt{Da(\mathbf{u}, \mathbf{w})[\mathbf{w}]} \|\mathbf{w}\|_{0,\Gamma_D} + \gamma\|\mathbf{w}\|_{0,\Gamma_D}^2 \\
&\geq (1 - \varepsilon)Da(\mathbf{u}, \mathbf{w})[\mathbf{w}] + \left(\gamma - \frac{C^2}{4\varepsilon}\right)\|\mathbf{w}\|_{0,\Gamma_D}^2.
\end{aligned}
$$

Here, we have subsequently used the Cauchy–Schwarz inequality, the inverse inequality (see Rüberg *et al.*, 2016, Appendix A)

$$C^2 Da(\mathbf{u}, \mathbf{w})[\mathbf{w}] \geq \|D\mathbf{t}(\mathbf{u})[\mathbf{w}]\|_{0,\Gamma_D}^2 \qquad (A.2)$$

and Young's inequality ($ab \leq \varepsilon a^2 + \frac{b^2}{4\varepsilon}$). One should notice that the constant C^2 in the inverse inequality (A.2) scales like the *bulk modulus* K denoted by

$$K = \frac{F_i}{3(1 - 2\nu)}. \qquad (A.3)$$

Since the parameters $E = 69$ GPa, $\nu = 0.334$ are chosen for aluminium stiffeners (The Engineering ToolBox, 2003, 2008), the bulk modulus is $K \approx 70$ GPa by (A.3). In addition, the material we considered in this work is compressible (Ciarlet, 1988) as $\nu < 0.5$.

Therefore, if we choose $\gamma > \frac{C^2}{4\varepsilon}$ and $\varepsilon < 1$, thus giving $\gamma > \frac{C^2}{4}$, it guarantees the positivity of $A(\mathbf{u}; \mathbf{w}, \mathbf{w})$ for any nonzero $\mathbf{w} \in V$. Hence, we should choose $\gamma > 10^{11}$ to guarantee solvability of the Newton system (A.1).

A.2   Equilibrium Equations in Two Dimensions

To construct the manufactured solution for numerical verification of the theoretical convergence order (see Section 9.3), we need to derive the strong form of the equilibrium equations of the minimisation problem. In two dimensions, the free energy functional to be minimised is

$$\mathcal{J}(u, \mathbf{Q}) = \int_\Omega \left(\frac{a_1}{2} u^2 + \frac{a_2}{3} u^3 + \frac{a_3}{4} u^4 \right.$$

$$+ B \left| \mathcal{D}^2 u + q^2 \left(\mathbf{Q} + \frac{\mathbf{I}_2}{2} \right) u \right|^2$$

$$\left. + \frac{K}{2} |\nabla \mathbf{Q}|^2 - l \left(\mathrm{tr}(\mathbf{Q}^2) \right) + l \left(\mathrm{tr}(\mathbf{Q}^2) \right)^2 \right),$$

with real parameters $a_1, a_2, a_3, B, q, K, l$. Note that $\mathbf{Q}$ is a symmetric and traceless 2×2 matrix and thus can be represented by two degrees of freedom (Q_{11}, Q_{12}) as given by (8.9). Then, we rewrite the above free energy in terms of variables (Q_{11}, Q_{12}, u) as follows,

$$\mathcal{J}(Q_{11}, Q_{12}, u)$$

$$= \int_\Omega \left(\frac{a_1}{2} u^2 + \frac{a_2}{3} u^3 + \frac{a_3}{4} u^4 \right.$$

$$+ B|\mathcal{D}^2 u|^2 + Bq^4 u^2 \left(2\left(Q_{11}^2 + Q_{12}^2\right) + \frac{1}{2} \right)$$

$$+ 2Bq^2 u \left(\left(Q_{11} + \frac{1}{2}\right) \partial_x^2 u + \left(-Q_{11} + \frac{1}{2}\right) \partial_y^2 u + 2Q_{12} \partial_x \partial_y u \right)$$

$$\left. + K |\nabla Q_{11}|^2 + K |\nabla Q_{12}|^2 - 2l \left(Q_{11}^2 + Q_{12}^2\right) + 4l \left(Q_{11}^2 + Q_{12}^2\right)^2 \right).$$
$$(A.4)$$

The admissible set for (Q_{11}, Q_{12}, u) based on (8.14) is denoted as

$$\tilde{\mathcal{A}}^s = \{u \in H^2(\Omega, \mathbb{R}), (Q_{11}, Q_{12}) \in H^1(\Omega, \mathbb{R}^2) \colon (Q_{11}, Q_{12}) = \mathbf{q}_b \text{ on } \partial\Omega\},$$

where $\mathbf{q}_b = (q_{b,1}, q_{b,2})^T$ is the prescribed Dirichlet boundary data arising from $\mathbf{Q}_b$.

Remark A.1. Note that the uniaxiality condition is not included in the admissible set here. This condition is beneficial for the variational analysis in Section 8.3.2, but enforcing the uniaxiality constraint strongly is not a trivial task (Borthagaray *et al.*, 2020). Instead, we weakly impose this constraint through the additional nematic bulk density $f_n^b(Q)$ in (8.12) which possesses a uniaxial minimiser by Majumdar and Zarnescu (2010, Proposition 15).

Remark A.2. Other choices of boundary data can be taken for (Q_{11}, Q_{12}); we choose Dirichlet boundary conditions for simplicity.

By taking the test functions $(p_1, p_2, v) \in H_0^1(\Omega) \times H_0^1(\Omega) \times H^2(\Omega)$ and using integration by parts, we derive the weak form of the Euler–Lagrange equations for the energy functional (A.4),

$$\mathcal{J}_{Q_{11}}(Q_{11}, Q_{12}, u; p_1) = \int_\Omega \left(4Bq^4 u^2 Q_{11} + 2Bq^2 u \left(\partial_x^2 u - \partial_y^2 u\right) \right.$$

$$\left. + 2K \Delta Q_{11} - 4l Q_{11} + 16l Q_{11} \left(Q_{11}^2 + Q_{12}^2\right) \right) p_1 = 0 \quad \forall p_1 \in H_0^1(\Omega),$$

$$\mathcal{J}_{Q_{12}}(Q_{11}, Q_{12}, u; p_2) = \int_\Omega \left(4Bq^4 u^2 Q_{12} + 4Bq^2 u(\partial_x \partial_y u) \right.$$

$$\left. + 2K\Delta Q_{12} - 4lQ_{12} + 16lQ_{12}\left(Q_{11}^2 + Q_{12}^2\right) \right) p_2 = 0 \quad \forall p_2 \in H_0^1(\Omega),$$

$$\mathcal{J}_u(Q_{11}, Q_{12}, u; v) = \int_\Omega \left(a_1 u + a_2 u^2 + a_3 u^3 + 2B\nabla \cdot (\nabla \cdot (\mathcal{D}^2 u)) \right.$$

$$+ Bq^4 \left(4\left(Q_{11}^2 + Q_{12}^2\right) + 1 \right) u$$

$$+ 2Bq^2 \left[(Q_{11} + 1/2)\partial_x^2 u + (-Q_{11} + 1/2)\partial_y^2 u + 2Q_{12}(\partial_x \partial_y u) \right]$$

$$+ 2Bq^2 \left[\partial_x^2(u(Q_{11} + 1/2)) + \partial_y^2(u(-Q_{11} + 1/2)) + 2\partial_x \partial_y(uQ_{12}) \right] \right) v$$

$$+ 2BG_{1,b}(u; v) + 2Bq^2 G_{2,b}(Q_{11}, Q_{12}, u; v) = 0 \quad \forall v \in H^2(\Omega),$$

where the boundary integrals $G_{1,b}$ and $G_{2,b}$ are of the form

$$G_{1,b}(u; v) = \int_{\partial\Omega} \nu \cdot (\mathcal{D}^2 u \cdot \nabla v) - \int_{\partial\Omega} ((\nabla \cdot (\mathcal{D}^2 u)) \cdot \nu) v$$

and

$$G_{2,b}(u, Q_{11}, Q_{12}; v)$$

$$= \int_{\partial\Omega} (-v (\partial_x(u(Q_{11} + 1/2))\nu_x) + (\partial_x v)u(Q_{11} + 1/2)\nu_x)$$

$$+ \int_{\partial\Omega} (-v (\partial_y(u(-Q_{11} + 1/2))\nu_y) + (\partial_y v)u(-Q_{11} + 1/2)\nu_y)$$

$$+ \int_{\partial\Omega} (-v (\partial_x(uQ_{12})\nu_y) + (\partial_y v)uQ_{12}\nu_x)$$

$$+ \int_{\partial\Omega} (-v (\partial_y(uQ_{12})\nu_x) + (\partial_x v)uQ_{12}\nu_y).$$

Therefore, the Euler–Lagrange equations for minimising the free energy (A.4) for $(Q_{11}, Q_{12}, u) \in \tilde{\mathcal{A}}^s$ are

$$4Bq^4 u^2 Q_{11} + 2Bq^2 u \left(\partial_x^2 u - \partial_y^2 u\right) - 2K\Delta Q_{11}$$

$$- 4lQ_{11} + 16lQ_{11}\left(Q_{11}^2 + Q_{12}^2\right) = 0,$$

$$4Bq^4 u^2 Q_{12} + 4Bq^2 u \left(\partial_x \partial_y u\right) - 2K\Delta Q_{12}$$

$$- 4lQ_{12} + 16lQ_{12}\left(Q_{11}^2 + Q_{12}^2\right) = 0,$$

$$a_1 u + a_2 u^2 + a_3 u^3 + 2B\nabla \cdot (\nabla \cdot (\mathcal{D}^2 u))$$

$$+ Bq^4 \left(4\left(Q_{11}^2 + Q_{12}^2\right) + 1\right) u + 2Bq^2(t_1 + t_2) = 0, \qquad \text{(A.5)}$$

subject to the boundary conditions

$$(Q_{11}, Q_{12}) = (q_{b,1}, q_{b,2}) \quad \text{on } \partial\Omega,$$

$$S_{bc}^1(u, q_{b,1}, q_{b,2}; v) = 0 \; \forall v \in H^2(\Omega) \quad \text{on } \partial\Omega,$$

where

$$t_1 := (Q_{11} + 1/2)\partial_x^2 u + (-Q_{11} + 1/2)\partial_y^2 u + 2Q_{12}\partial_x\partial_y u,$$

$$t_2 := \partial_x^2 \left(u\left(Q_{11} + 1/2\right)\right) + \partial_y^2(u(-Q_{11} + 1/2)) + 2\partial_x\partial_y(uQ_{12}),$$

$$S_{bc}^1(u, q_{b,1}, q_{b,2}; v) := G_{1,b}(u; v) + q^2 G_{2,b}(u, q_{b,1}, q_{b,2}; v).$$

Equations (A.5) are used for the numerical verification of the theoretical convergence rates derived in Chapter 9. Here, we will not derive the equilibrium equations for three-dimensional problems due to their complicated form with six coupled degrees of freedom $(Q_{11}, Q_{12}, Q_{13}, Q_{22}, Q_{23}, u)$.

Bibliography

Adler, J. H., Atherton, T. J., Benson, T., Emerson, D. B., and MacLachlan, S. P. (2015a). Energy minimization for liquid crystal equilibrium with electric and flexoelectric effects, *SIAM J. Sci. Comput.* **37**, 5, pp. S157–S176.

Adler, J. H., Atherton, T. J., Emerson, D. B., and Maclachlan, S. P. (2015b). An energy-minimization finite element approach for the Frank–Oseen model of nematic liquid crystals, *SIAM J. Numer. Anal.* **53**, 5, pp. 2226–2254.

Adler, J. H., Atherton, T. J., Emerson, D. B., and Maclachlan, S. P. (2016). Constrained optimization for liquid crystal equilibria, *SIAM J. Sci. Comput.* **38**, 1, pp. 50–76.

Ahrens, J., Geveci, B., and Law, C. (2005). Paraview: an end-user tool for large-data visualization, in C. D. Hansen and C. R. Johnson (eds.), *Visualization Handbook* (Butterworth-Heinemann, Burlington), pp. 717–731, ISBN 978-0-12-387582-2, doi:10.1016/B978-012387582-2/50038-1.

Amestoy, P. R., Duff, I., and L'Excellent, J.-Y. (2000). Multifrontal parallel distributed symmetric and unsymmetric solvers, *Comput. Methods Appl. Mech. Eng.* **184**, 2–4, pp. 501–520.

Argyris, J. H., Fried, I., and Scharpf, D. W. (1968). The TUBA family of plate elements for the matrix displacement method, *Aeronaut. J.* **72**, pp. 701–709.

Arnold, D. N., Falk, R. S., and Winther, R. (2000). Multigrid in H(div) and H(curl), *Numer. Math.* **85**, pp. 197–217.

Averill, B. and Eldredge, P. (2011). *General Chemistry: Principles, Patterns, and Applications*, Chap. 11 (Saylor Academy, Minneapolis).

Balay, S., Abhyankar, S., Adams, M. F., Brown, J., Brune, P., Buschelman, K., Dalcin, L., Eijkhout, V., Gropp, W. D., Kaushik, D., Knepley, M. G., McInnes, L. C., Rupp, K., Smith, B. F., and Zhang, H. (2018). PETSc users manual, Tech. Rep. ANL-95/11 — Revision 3.9, Argonne National Laboratory.

Ball, J. M. (2017). Mathematics and liquid crystals, *Mol. Cryst. Liq. Cryst.* **647**, 1, pp. 1–27.

Ball, J. M. and Bedford, S. J. (2015). Discontinuous order parameters in liquid crystal theories, *Mol. Cryst. Liq. Cryst.* **612**, 1, pp. 1–23.

Ball, J. M. and Zarnescu, A. (2008). Orientable and non-orientable line field models for uniaxial nematic liquid crystals, *Mol. Cryst. Liq. Cryst.* **495**, pp. 221/[573]–233/[585].

Bedford, S. J. (2014). *Calculus of Variations and Its Application to Liquid Crystals*, Ph.d thesis, University of Oxford.

Benzi, M., Golub, G. H., and Liesen, J. (2005). Numerical solution of saddle point problems, *Acta Numer.* **14**, pp. 1–137.

Benzi, M. and Olshanskii, M. A. (2006). An augmented Lagrangian-based approach to the Oseen problem, *SIAM J. Sci. Comput.* **28**, 6, pp. 2095–2113.

Birkisson, Á. (2013). *Numerical Solution of Nonlinear Boundary Value Problems for Ordinary Differential Equations in the Continuous Framework*, Ph.d thesis, University of Oxford.

Biscari, P., Calderer, M. C., and Terentjev, E. (2007). Landau–de Gennes theory of isotropic-nematic-smectic liquid crystal transitions, *Phys. Rev. E* **75**, pp. 051707-1–051707-11.

Bisht, K., Banerjee, V., Milewski, P., and Majumdar, A. (2019). Magnetic nanoparticles in a nematic channel: a one-dimensional study, *Phys. Rev. E* **100**, 012703, pp. 012703-1–012703-9.

Blum, H. and Bonn, R. R. (1980). On the boundary value problem of the biharmonic operator on domains with angular corners, *Math. Mech. in the Appli. Sci.* **2**, pp. 556–581.

Borthagaray, J. P., Nochetto, R. H., and Walker, S. W. (2020). A structure-preserving FEM for the uniaxially constrained **q**-tensor model of nematic liquid crystals, *Numer. Math.* **145**, 4, pp. 837–881, doi:10.1007/s00211-020-01133-z.

Borzí, A. (2002). Introduction to multigrid methods, Tech. rep., Institut für Mathematik und Wissenschaftliches Rechnen, https://www.mathematik.uni-wuerzburg.de/~borzi/mgintro.pdf.

Bower, A. F. (2009). *Applied Mechanics of Solids* (CRC Press), ISBN 9781439802472, http://solidmechanics.org/contents.php.

Braess, D. and Schumaker, L. (1997). *Finite Elements: Theory, Fast Solvers, and Applications in Solid Mechanics* (Cambridge University Press), ISBN 9780521581875, https://books.google.co.uk/books?id=GM1ltAEACAAJ.

Braides, A. (2006). A handbook of Γ-convergence, in *Handbook of Differential Equations: Stationary Partial Differential Equations*, Vol. 3 (Elsevier, North-Holland, Amsterdam), pp. 101–213.

Brenner, S. C. (2003). Poincaré-Friedrichs inequalities for piecewise H^1 functions, *SIAM J. Numer. Anal.* **41**, pp. 306–324.

Brenner, S. C. (2011). c^0 interior penalty methods, in J. Blowey and M. Jensen (eds.), *Frontiers in Numerical Analysis — Durham 2010. Lecture Notes in Computational Science and Engineering*, Vol. 85 (Springer, Berlin, Heidelberg), doi:https://doi.org/10.1007/978-3-642-23914-4_2.

Brenner, S. C., Gudi, T., and Sung, L. (2010). A weakly over-penalized symmetric interior penalty method for the biharmonic problem, *Electon. Trans. Numer. Anal.* **37**, pp. 214–238.

Brenner, S. C. and Scott, L. R. (2008). *The Mathematical Theory of Finite Element Methods, Texts in Applied Mathematics*, Vol. 15, 3rd edn. (Springer, New York).

Brenner, S. C. and Sung, L. (2005). c^0 interior penalty methods for fourth order elliptic boundary value problems on polygonal domains, *J. Sci. Comput.* **22**, pp. 83–118.

Brenner, S. C. and Sung, L. (2008). A weakly over-penalized symmetric interior penalty method, *Electon. Trans. Numer. Anal.* **30**, pp. 107–127.

Briggs, W., Henson, V., and McCormick, S. (2000). *A Multigrid Tutorial*, 2nd edn. (Society for Industrial and Applied Mathematics), Philadelphia, PA USA, doi:10.1137/1.9780898719505.

Brochard, F. and de Gennes, P. G. (1970). Theory of magnetic suspensions in liquid crystals, *J. Phys. France* **31**, 7, pp. 691–708.

Brown, K. M. and Gearhart, W. B. (1971). Deflation techniques for the calculation of further solutions of a nonlinear system, *Numerische Mathematik* **16**, 4, pp. 334–342.

Brune, P. R., Knepley, M. G., Smith, B. F., and Tu, X. (2015). Composing scalable nonlinear algebraic solvers, *SIAM Rev.* **57**, 4, pp. 535–565, doi: 10.1137/130936725.

Burylov, S. V. and Raikher, Y. L. (1995). Macroscopic properties of ferronematics caused by orientational interactions on the particle surfaces. I. Extended continuum model, *Mol. Cryst. Liq. Cryst. Sci. Technol. Sect. A* **258**, pp. 107–122.

Calderer, M. C., DeSimone, A., Golovaty, D., and Panchenko, A. (2014). An effective model for nematic liquid crystal composites with ferromagnetic inclusions, *SIAM J. Appl. Math.* **74**, pp. 237–262.

Calderer, M. C. and Palffy-Muhoray, P. (2000). Ericksen's bar and modeling of the smectic A-nematic phase transition, *SIAM J. Appl. Math.* **60**, 3, pp. 1073–1098.

Canevari, G., Majumdar, A., and Spicer, A. (2019). Order reconstruction for nematics on squares and hexagons: a Landau–de Gennes study, *SIAM J. Appl. Math.* **77**, pp. 267–293.

Carlsson, T., Stewart, I. W., and Leslie, F. M. (1991). Theoretical studies of smectic C liquid crystals confined in a wedge. Stability considerations and Frederiks transitions, *Liq. Cryst.* **9**, 5, pp. 661–678.

Chandrasekhar, S. (1992). *Liquid Crystals*, 2nd edn. (Cambridge University Press, Cambridge, UK).

Cheng, X., Han, W., and Huang, H. (2000). Some mixed finite element methods for the biharmonic equation, *J. Comp. Appl. Math.* **126**, pp. 91–109.

Ciarlet, P. (2002). *The Finite Element Method for Elliptic Problems* (Society for Industrial and Applied Mathematics).

Ciarlet, P. G. (1978). *The Finite Element for Elliptic Problems* (North-Holland, Amsterdam, New York, Oxford).

Ciarlet, P. G. (1988). *Mathematical Elasticity, Vol. I: Three-dimensional Elasticity*, Studies in Mathematics and its Applications (North-Holland), ISBN 0-444-70259-8.

Clément, P. (1975). Approximation by finite element functions using local regularization, *Rev. Française Automat. Informat. Recherche Opérationnelle Sér. Rouge Anal. Numér.* **9**, R-2, pp. 77–84.

Dalby, J., Farrell, P. E., Majumdar, A., and Xia, J. (2022). One-dimensional ferronematics in a channel: order reconstruction, bifurcations and multistability, *SIAM J. Appl. Math.* **82**, 2, pp. 694–719.

Davis, T. A. (1994). *Finite Element Analysis of the Landau–de Gennes Minimization Problem for Liquid Crystals in Confinement*, Ph.d thesis, Kent State University.

Davis, T. A. and Gartland, J. E. C. (1998). Finite element analysis of the Landau-de Gennes minimization problem for liquid crystals, *SIAM J. Numer. Anal.* **35**, pp. 336–362.

de Gennes, P. G. (1969). Phenomenology of short-range-order effects in the isotropic phase of nematic materials, *Phys. Lett.* **30A**, 8, pp. 454–455.

de Gennes, P. G. (1972). An analogy between superconductors and smectic A, *Solid State Commun.* **10**, pp. 753–756.

de Gennes, P. G. (1973). Some remarks on the polymorphism of smectics, *Mol. Cryst. Liq. Cryst.* **21**, pp. 49–76.

de Gennes, P. G. (1974). *The Physics of Liquid Crystals* (Oxford University Press, Oxford).

Dener, A., Denchfield, A., Suh, H., Munson, T., Sarich, J., Wild, S., Benson, S., and McInnes, F. C. (2020). Toolkit for advanced optimization (tao) users manual, Tech. Rep. ANL/MCS-TM-322 - Revision 3.14, Argonne National Laboratory.

Deuflhard, P. (2011). *Newton Methods for Nonlinear Problems: Affine Invariance and Adaptive Algorithms* (Springer Publishing Company, Incorporated), ISBN 364223898X, 9783642238987.

Dunmur, D. and Sluckin, T. (2011). *Soap, Science, and Flat-Screen TVs* (Oxford University Press, New York, United States).

E, W. (1997). Nonlinear continuum theory of smectic-a liquid crystals, *Arch. Rational Mech. Anal.* **137**, pp. 159–175.

Elman, H. C., Silvester, D., and Wathen, A. J. (2014). *Finite Elements and Fast Iterative Solvers: With Applications in Incompressible Fluid Dynamics*, 2nd edn. (Oxford University Press, Oxford, UK), ISBN 9780199678792.

Embar, A., Dolbow, J., and Harari, I. (2010). Imposing Dirichlet boundary conditions with Nitsche's method and spline-based finite elements, *International J. Numer. Meth. Eng.* **83**, pp. 877–898.

Emerson, D. B. (2015). *Advanced Discretizations and Multigrid Methods for Liquid Crystal Configurations*, Ph.d thesis, Tufts University.

Emerson, D. B., Farrell, P. E., Adler, J. H., MacLachlan, S. P., and Atherton, T. J. (2018). Computing equilibrium states of cholesteric liquid crystals in elliptical channels with deflation algorithms, *Liq. Cryst.* **45**, 3, pp. 341–350.

Engel, G., Garikipati, K., Hughes, T. J. R., Larson, M. G., Mazzei, L., and Taylor, R. L. (2002). Continuous/discontinuous finite element approximations of fourth-order elliptic problems in structural and continuum mechanics with applications to thin beams and plates, and strain gradient elasticity, *Comput. Meth. Appl. Mech. Engrg.* **191**, pp. 3669–3750.

Evans, L. C. (2010). *Partial Differential Equations*, Vol. 19 of Graduate Studies in Mathematics, 2nd edn. (American Mathematical Society, Providence, RI).

Falgout, R., Yang, U., and Li, R. (2018). Multigrid and Multilevel Methods, https://fastmath-scidac.llnl.gov/research/multigrid-and-multilevel-methods.html.

Farrell, P. E. (2017). Defcon, Https://bitbucket.org/pefarrell/defcon/src/master/.

Farrell, P. E., Beentjes, C. H. L., and Birkisson, Á. (2016). The computation of disconnected bifurcation diagrams, preprint (2016), arXiv:1603.00809v1 [math.NA].

Farrell, P. E., Birkisson, Á., and Funke, S. W. (2015). Deflation techniques for finding distinct solutions of nonlinear partial differential equations, *SIAM J. Sci. Comput.* **37**, 4, pp. A2026–A2045.

Farrell, P. E., Knepley, M. G., Wechsung, F., and Mitchell, L. (2021). Pcpatch: software for the topological construction of multigrid relaxation methods, *ACM Trans. Math. Softw.* arXiv:1912.08516, https://arXiv.org/abs/1912.08516, in press.

Farrell, P. E., Mitchell, L., and Wechsung, F. (2019). An augmented Lagrangian preconditioner for the 3D stationary incompressible Navier–Stokes equations at high Reynolds number, *SIAM J. Sci. Comput.* **41**, pp. A3073–A3096.

Firedrake-Zenodo (2020). Software used in 'Augmented Lagrangian preconditioners for the Oseen–Frank model of nematic and cholesteric liquid crystals', https://doi.org/10.5281/zenodo.4249051.

Firedrake-Zenodo (2021a). Software used in 'One-dimensional ferronematics in a channel — order reconstruction, bifurcations and multistability', https://doi.org/10.5281/zenodo.4449535.

Firedrake-Zenodo (2021b). Software used in 'Structural Transitions in Geometrically Frustrated Smectics', https://doi.org/10.5281/zenodo.4441123.

Fortin, M. and Glowinski, R. (1983). *Augmented Lagrangian Methods: Applications to the Numerical Solution of Boundary-Value Problems, Studies in Mathematics and Its Applications*, Vol. 15 (Elsevier Science Ltd., North-Holland - Amsterdam, New York, Oxford).

Frank, F. C. (1958). Liquid crystals, *Faraday Discuss.* **25**, pp. 19–28.

Freund, J. and Stenberg, R. (1995). On weakly imposed boundary conditions for second order problems, in *Proceedings of the Ninth International Conference on Finite Elements in Fluids*, pp. 327–336.

Friedel, V. (1922). Les états mésomorphes de la matiére, *Ann. Phys.* **18**, pp. 273–474.

Geuzaine, C. and Remacle, J.-F. (2009). Gmsh: a three-dimensional finite element mesh generator with built-in pre- and post-processing facilities, *Int. J. Numer. Meth. Eng.* **79**, 11, pp. 1309–1331.

Giaquinta, M. (1983). *Multiple Integrals in the Calculus of Variations and Nonlinear Elliptic Systems* (Princeton University Press, Princeton, New Jersey).

Girault, V. and Raviart, P. A. (2011). *Finite Element Methods for Navier–Stokes Equations: Theory and Algorithms*, 1st edn. (Springer, Berlin, Heidelberg).

Glowinski, R. and Le Tallec, P. (1989). *Augmented Lagrangian Methods for the Solution of Variational Problems*, Chap. 3, Studies in Applied Mathematics (SIAM), ISBN 978-0-89871-230-8, pp. 45–121.

Glowinski, R., Lin, P., and Pan, X. B. (2003). An operator-splitting method for a liquid crystal model, *Comput. Phys. Commun.* **152**, 3, pp. 242–252.

Grisvard, P. (1985). *Elliptic Problems in Nonsmooth Domains*, 1st edn. (Pitman Advanced Publishing Program), ISBN 0-273-08647-2.

Guyan, R. J. (1965). Reduction of stiffness and mass matrices, *AIAA J.* **3**, 2, p. 380.

Hager, W. W. (1989). Updating the inverse of a matrix, *SIAM Rev.* **31**, 2, pp. 221–239.

Halperin, B. I. and Lubensky, T. C. (1974). On the analogy between smectic A liquid crystals and superconductors, *Solid State Commun.* **14**, pp. 997–1001.

Han, J., Luo, Y., Wang, W., Zhang, P., and Zhang, Z. (2015). From microscopic theory to macroscopic theory: a systematic study on modeling for liquid crystals, *Arch. Rational Mech. Anal.* **215**, pp. 741–809.

He, X., Vuik, C., and Klaij, C. M. (2018). Combining the augmented Lagrangian preconditioner with the simple Schur complement approximation, *SIAM J. Sci. Comput.* **40**, 3, pp. A1362–A1385.

Heister, T. and Rapin, G. (2012). Efficient augmented Lagrangian-type preconditioning for the Oseen problem using Grad-Div stabilization, *Int. J. Numer. Meth. Fl.* **71**, 1, pp. 118–134.

Henson, V. E. (2003). Multigrid methods for nonlinear problems: an overview, in *International Society for Optical Engineering 15th Annual Conference on Electronic Imaging* (Santa Clara, California), pp. 1–15.

Hernandez, V., Roman, J. E., and Vidal, V. (2005). SLEPc: A scalable and flexible toolkit for the solution of eigenvalue problems, *ACM Trans. Math. Softw.* **31**, 3, pp. 351–362, doi:hernandez2005.

Hu, Q., Tai, X., and Winther, R. (2009). A saddle point approach to the computation of harmonic maps, *SIAM J. Numer. Anal.* **47**, 2, pp. 1500–1523.

Huang, H. and Asay, J. R. (2005). Compressive strength measurements in aluminum for shock compression over the stress range of 4-22 GPa, *J. Appl. Phys.* **033524**, pp. 1–17.

Irons, B. (1965). Structural eigenvalue problems: elimination of unwanted variables, *AIAA J.* **3**, 5, pp. 961–962.

John, V., Linke, A., Merdon, C., Neilan, M., and Rebholz, L. (2017). On the divergence constraint in mixed finite element methods for incompressible flows, *SIAM Rev.* **59**, 3, pp. 492–544.

Kantorovich, L. (1948). On Newton's method for functional equations, *Dokl. Akad. Nauk SSSR* **59**, pp. 1237–1249.

Kesavan, S. (1989). *Topics in Functional Analysis and Applications* (John Wiley & Sons, New York).

Kim, Y. H., Yoon, D. K., Jeong, H. S., Lavrentovich, O. D., and Jung, H.-T. (2011). Smectic liquid crystal defects for self-assembling of building blocks and their lithographic applications, *Adv. Funct. Mater.* **21**, 4, pp. 610–627.

Knabner, P. and Angermann, L. (2000). *Numerik Partieller Differentialgleichungen* (Springer-Verlag: Berlin, Heidelberg, New York).

Lacaze, E., Michel, J.-P., Alba, M., and Goldmann, M. (2007). Planar anchoring and surface melting in the smectic-A phase, *Phys. Rev. E* **76**, 4, p. 041702, doi:10.1103/physreve.76.041702.

Lagerwall, J. P. F. and Scalia, G. (2012). A new era for liquid crystal research: Applications of liquid crystals in soft matter nano-, bio- and microtechnology, *Curr. Appl. Phys.* **12**, pp. 1387–1412.

Lamy, X. (2014). Bifurcation analysis in a frustrated nematic cell, *J. Nonlin. Sci.* **24**, pp. 1197–1230.

Lee, Y., Wu, J., Xu, J., and Zikatanov, L. (2007). Robust subspace correction methods for nearly singular systems, *Math. Mod. Meth. Appl. S.* **17**, 11, pp. 1937–1963.

Lelidis, I., Blanc, C., and Klèman, M. (2006). Optical and confocal microscopy observations of screw dislocations in smectic-A liquid crystals, *Phys. Rev. E* **74**, 051710, pp. 1–5.

Lin, F. (1989). Nonlinear theory of defects in nematic liquid crystals; phase transition and flow phenomena, *Commun. Pur. Appl. Math.* **42**, 6, pp. 789–814.

Lin, P. and Richter, T. (2007). An adaptive homotopy multi-grid method for molecule orientations of high dimensional liquid crystals, *J. Comput. Phys.* **225**, 2, pp. 2069–2082.

Lin, P. and Tai, X. (2014). An Augmented Lagrangian Method for the Microstructure of a Liquid Crystal Model, in W. Fitzgibbon, Y. Kuznetsov, P. Neittaanmäki, and O. Pironneau (eds.), *Modeling, Simulation and Optimization for Science and Technology*, Vol. 34, chap. 7 (Springer, Dordrecht), ISBN 978-94-017-9053-6, pp. 123–137.

Linhananta, A. and Sullivan, D. E. (1991). Phenomenological theory of smectic — a liquid crystals, *Phys. Rev. A* **44**, 12, pp. 8189–8197.

Lu, K., Augarde, C. E., Coombs, W. M., and Hu, Z. (2019). Weak impositions of Dirichlet boundary conditions in solid mechanics: a critique of current approaches and extension to partially prescribed boundaries, *Comput. Meth. Appl. Mech. Eng.* **348**, pp. 632–659.

Maity, R. R., Majumdar, A., and Nataraj, N. (2020). Discontinuous Galerkin finite element methods for the Landau–de Gennes minimization problem of liquid crystals, *IMA J. Numer. Anal.* **00**, pp. 1–34.

Majumdar, A. and Zarnescu, A. (2010). Landau-de Gennes theory of nematic liquid crystals: the Oseen–Frank limit and beyond, *Arch. Ration. Mech. Anal.* **196**, pp. 227–280.

Mei, S. and Zhang, P. (2015). On a molecular based Q-tensor model for liquid crystals with density variations, *Multiscale Model. Simul.* **13**, 3, pp. 977–1000.

Mertelj, A., Lisjak, D., Drofenik, M., and Čopič, M. (2013). Ferromagnetism in suspensions of magnetic platelets in liquid crystals, *Nature* **504**, pp. 237–241.

Michel, J.-P., Lacaze, E., Alba, M., de Boissieu, M., Gailhanou, M., and Goldmann, M. (2004). Optical gratings formed in thin smectic films frustrated on a single crystalline substrate, *Phys. Rev. E* **70**, 011709, pp. 1011709-1–011709-12.

Michel, J.-P., Lacaze, E., and Goldmann, M. (2006). Structure of smectic defect cores: X-ray study of 8CB liquid crystal ultrathin films, *Phys. Rev. Lett.* **96**, 2, p. 027803, doi:10.1103/physrevlett.96.027803.

Mottram, N. J. and Newton, C. J. P. (2014). Introduction to Q-tensor theory, arXiv:1409.3502, https://arxiv.org/abs/1409.3542.

Nitsche, J. (1971). Über ein Variationsprinzip zur Lösung von Dirichlet-Problemen bei Verwendung von Teilräumen, die keinen Randbedingungen unterworfen sind, *Abhandlungen aus dem Mathematischen Seminar der Universität Hamburg* **36**, 1, pp. 9–15.

Nocedal, J. and Wright, S. J. (1999). *Numerical Optimisation* (Springer), ISBN 0387987932.

Nochetto, R. H., Walker, S. W., and Zhang, W. (2017). A finite element method for liquid crystals with variable degree of orientation, *SIAM J. Numer. Anal.* **55**, pp. 1357–1386.

Ogawa, H. and Uchida, N. (2006). Numerical simulation of the twist-grain-boundary phase of chiral liquid crystals, *Phys. Rev. E* **73**, pp. 060701-1–060701-4.

Olshanskii, M. A. (2002). A low order Galerkin finite element method for the Navier–Stokes equations of steady incompressible flow: a stabilization issue and iterative methods, *Comput. Meth. Appl. M.* **191**, 47, pp. 5515–5536.

Oseen, C. W. (1933). The theory of liquid crystals, *Trans. Faraday Soc.* **29**, 140, pp. 883–899.

Pevnyi, M. Y., Selinger, J., and Sluckin, T. J. (2014). Modeling smectic layers in confined geometries: order parameter and defects, *Phys. Rev. E*, **90**, 032507, pp. 1–8.

Polyak, V. T. and Tret'yakov, N. V. (1974). The method of penalty estimates for conditional extremum problems, *USSR Comput. Math. Math. Phys.* **13**, 1, pp. 42–58.

Poniewierski, A. and Sluckin, T. J. (1991). Phase diagram for a system of hard spherocylinders, *Phys. Rev. A* **43**, 12, pp. 6837–6842.

Rathgeber, F., Ham, D. A., Mitchell, L., Lange, M., Luporini, F., McRae, A. T. T., Bercea, G. T., Markall, G. R., and Kelly, P. H. J. (2017). Firedrake: automating the finite element method by composing abstractions, *ACM T. Math. Softw.* **43**, 3, pp. 1–27.

Rault, J., Cladis, P. E., and Burger, J. P. (1970). Ferronematics, *Phys. Lett. A* **32**, pp. 199–200.

Reinitzer, F. (1888). Beiträge zur Kenntnis des Cholesterins, *Monatsh. Chem.* **9**, pp. 421–441.

Rüberg, T., Cirak, F., and Aznar, J. M. G. (2016). An unstructured immersed finite element method for nonlinear solid mechanics, *Adv. Model. Simul. Eng. Sci.* **3**, 22, pp. 1–28.

Saad, Y. (1993). A flexible inner-outer preconditioned GMRES algorithm, *SIAM J. Sci. Comput.* **14**, 2, pp. 461–469.

Saad, Y. and Schultz, M. (1986). GMRES: a generalized minimal residual algorithm for solving nonsymmetric linear systems, *SIAM J. Sci. Statist. Comput.* **7**, 3, pp. 856–869.

Santangelo, C. D. and Kamien, R. D. (2007). Triply periodic smectic liquid crystals, *Phys. Rev. E* **75**, 011702, pp. 011702-1–011702-12.

Schöberl, J. (1999a). Multigrid methods for a parameter dependent problem in primal variables, *Numer. Math.* **84**, 1, pp. 97–119.

Schöberl, J. (1999b). *Robust Multigrid Methods for Parameter Dependent Problems*, Ph.d thesis, Johannes Kepler University Linz.

Scholtz, R. (1978). A mixed method for fourth-order problems using linear finite elements, *RAIRO Numer. Anal.* **15**, pp. 85–90.

SciencebyDegrees (2018). An introduction to liquid crystals, https://sciencebyde grees.com/2018/08/10/liquid-crystals/.

Shi, B., Han, Y., and Zhang, L. (2022). Nematic liquid crystals in a rectangular confinement: solution landscape and bifurcation, *SIAM J. Appl. Math.* **82**, 5, pp. 1808–1828.

Silvester, D. and Wathen, A. J. (1994). Fast iterative solution of stabilised Stokes systems. Part II: using general block preconditioners, *SIAM J. Numer. Anal.* **31**, pp. 1352–1367.

Simo, J. C. and Hughes, T. J. R. (2000). *Computational Inelasticity* (Springer-Verlag New York), ISBN 978-0-387-22763-4.

Stewart, G. W. (2002). A Krylov–Schur algorithm for large eigenproblems, *SIAM J. Matrix Anal. Appl.* **23**, 3, pp. 601–614, doi:10.1137/S0895479800371529.

Stewart, I. W. (2004). *The Static and Dynamic Continuum Theory of Liquid Crystals: A Mathematical Introduction* (CPC Press), ISBN 9780748408962.

Süli, E. and Mozolevski, I. (2007). hp-version interior penalty DGFEMs for the biharmonic equation, *Comput. Meth. Appl. Mech. Engrg.* **196**, pp. 1851–1863.

The Engineering ToolBox (2003). Young's modulus — tensile and yield strength for common materials, https://www.engineeringtoolbox.com/young-modu lus-d_417.html.

The Engineering ToolBox (2008). Poisson's ratio, https://www.engineeringtoolb ox.com/poissons-ratio-d_1224.html.

Vanka, S. P. (1986). Block-implicit multigrid calculation of two-dimensional recirculating flows, *Comput. Meth. Appl. M.* **59**, 1, pp. 29–48.

Wang, W., Zhang, L., and Zhang, P. (2021). Modeling and computation of liquid crystals, *Acta Numer.* **30**, pp. 765–851.

Wang, Y., Canevari, G., and Majumdar, A. (2019). Order reconstruction for nematics on squares with isotropic inclusions: a Landau–de Gennes study, *SIAM J. Appl. Math.* **79**, pp. 1314–1340.

Wathen, A. J. and Silvester, D. (1991). Fast iterative solution of stabilised Stokes systems. Part I: Using simple diagonal preconditioners, *SIAM J. Numer. Anal.* **30**, 3, pp. 630–649.

Wilkinson, J. H. (1994). *Rounding Errors in Algebraic Processes* (Dover Publications, New York, NY, USA), ISBN 0486679993.

Williams, C. and Kléman, M. (1975). Dislocations, grain boundaries and focal conics in smectics A, *J. Phys. Colloq.* **36**, C1, pp. C1-315–C1-320.

Wittmann, R., Cortes, L. B. G., Löwen, H., and Aarts, D. G. A. L. (2021). Particle-resolved topological defects of smectic colloidal liquid crystals in extreme confinement, *Nat. Comm.* **12**, 623, pp. 1–20.

Xia, J. (2020). ALpaper-numerics, https://doi.org/10.5281/zenodo.4257094.

Xia, J. (2021a). Ferronematics-numerics, doi:10.5281/zenodo.4616745, https://doi.org/10.5281/zenodo.4616745.

Xia, J. (2021b). Smectic-A numerics, https://doi.org/10.5281/zenodo.4607849.

Xia, J. (2021c). Structural transitions in geometrically frustrated smectics (supplementary materials), https://www.youtube.com/playlist?list=PLr8tas_d t-wwd81QWCyNZe51L7NckSfwo.

Xia, J. and Farrell, P. E. (2023). Variational and numerical analysis of a Q-tensor model for smectic-A liquid crystals, *ESAIM Math. Model. Numer. Anal.* **57**, 2, pp. 1–25.

Xia, J., Farrell, P. E., and Castro, S. G. P. (2020). Nonlinear bifurcation analysis of stiffener profiles via deflation techniques, *Thin-Walled Struct.* **149**, 106662, pp. 1–25.

Xia, J., Farrell, P. E., and Wechsung, F. (2021a). Augmented lagrangian preconditioners for oseen–frank models in nematic and cholesteric liquid crystals, *BIT Numer. Math.*, pp. 1–38, https://doi.org/10.1007/s10543-020-00838-9.

Xia, J., MacLachlan, S., Atherton, T. J., and Farrell, P. E. (2021b). Structural landscapes on geometrically frustrated smectics, *Phys. Rev. Lett.* **126**, pp. 177801-1–1779801-6.

Xu, J. (1992). Iterative methods by space decomposition and subspace correction, *SIAM Rev.* **34**, 4, pp. 581–613.

Yao, X., Zhang, L., and Chen, J. (2022). Defect patterns of two-dimensional nematic liquid crystals in confinement, *Phys. Rev. E* **105**, 044704, pp. 1–7.

Yin, J., Wang, Y., Chen, J., Zhang, P., and Zhang, L. (2020). Construction of a pathway map on a complicated energy landscape, *Phys. Rev. Lett.* **124**, 090601, pp. 1–7.

Zappone, B. and Lacaze, E. (2008). Surface-frustrated periodic textures of smectic-A liquid crystals on crystalline surfaces, *Phys. Rev. E*, 061704, pp. 061704-1–061704-9.

Zappone, B., Mamuk, A. E., Gryn, I., Arima, V., Zizzari, A., Bartolino, R., Lacaze, E., and Petschek, R. (2020). Analogy between periodic patterns in thin smectic liquid crystal films and the intermediate state of superconductors, *Proc. Natl. Acad. Sci. USA* **117**, 30, pp. 17643–17649.

Zhang, S. (2009). A family of 3D continuously differentiable finite elements on tetrahedral grids, *Appl. Numer. Math.* **59**, pp. 219–233.

Index

a priori error estimates, 135, 137, 146, 172, 196
additive Schwarz preconditioner, 46, 47
admissible space, 72, 76, 129
aircraft stiffener profiles, 96
ALLU, 55, 56
ALMG-PBJ, 53, 57
ALMG-STAR, 53, 57
arc-length continuation, 89
aspect ratio, 189
asymptotic expansions, 113, 115
augmented discrete system, 24
augmented Lagrangian, 13, 20, 195
average operator, 149

biaxial, 68
biaxiality parameter, 85
bifurcation analysis, 97, 196
bifurcation diagram, 114, 116, 182
block Gaussian elimination, 13
"bookshelf" structure, 183
bootstrapping argument, 166
boundary value problem, 93
boundedness, 19, 28, 141, 152, 167
branch cut, 67, 124
branch switching, 89
broken norm, 148
broken Sobolev space, 140, 148

Brouwer's fixed point theorem, 159, 163
bubble space, 16
bulk constants, 69
bulk energy, 68, 78, 129
bulk minimisers, 75
bulk modulus, 204

C^0 interior penalty methods, 146, 198
Cauchy–Schwarz inequality, 24, 31, 144, 152
"chevron" structure, 198
cholesteric liquid crystals, 2
coarse-grid correction, 37, 38
coercivity, 17, 21, 153
confined geometries, 200
conforming finite elements, 15
consistency, 149
contraction, 162
convergence rate, 74, 171, 174, 176, 196

de Gennes theory, 121, 122
de Rham complexes, 197
decoupled case, 128
defect free, 179, 180
deflated continuation, 90
deflation, 88, 89
deflation operator, 90
deformation gradient, 91

deformed configuration, 91
Dirichlet boundary conditions, 72
domain wall, 77, 196
dual pairing, 148
dual solution, 142
Dupin cyclide, 184, 188

eccentricity, 188, 189
elastic constants, 10, 69
elastic energy density, 68
elliptic regularity, 150, 154
energy landscape, 191
enrichment estimates, 156
enrichment operator, 153
equal-constant approximation, 10, 30
equilibrium equations, 14, 20, 136,
    204
Ericksen's inequality, 12
Euclidean norm, 35
Euler–Lagrange equations, 72, 113,
    205, 206
exact solutions, 170
existence of minimisers, 129, 130

F-cycle, 39
ferronematics, 70, 71
finite elements, 30, 135, 196
first Piola–Kirchhoff stress tensor, 92
focal conic domain, 179, 184
fourth-order elasticity tensor, 95
fourth-order PDE, 135, 147
Fréchet derivative, 54
Frobenius inner product, 82
full approximation scheme, 40–42

Γ-convergence, 78
Galerkin subspace, 46
Gauss–Seidel, 36

H^1 error estimates, 139
Hölder spaces, 131
Hölder inequality, 31, 145
half-charge defects, 126
harmonic map, 10
head-to-tail symmetry, 3, 124
Hessian matrix, 99, 180

homeotropic anchoring, 189
Hsieh–Clough–Tocher finite elements,
    153

in-plane radial configuration, 185
inconsistent discrete form, 169
inf-sup condition, 19, 28, 140, 151,
    157
initial guess, 185
instability, 78
integration by parts, 143
interpolation estimates, 141, 156, 157,
    168
interpolation operator, 141
inverse inequality, 51, 204
isotropic, 68
iterative methods, 35, 37

Jacobi method, 36

kernel, 33, 43, 45, 196, 197
kernel capturing, 44, 49, 51
kernel capturing relaxation, 43
Kronecker delta, 95

L^2 error estimates, 142, 143, 164,
    166
Lagrange multiplier, 14
Lagrangian, 20, 27
Lagrangian strain tensor, 91
Lamé parameters, 91
Landau–de Gennes, 67, 68, 195
Laplace equations, 74
linear dual problem, 164
liquid crystals, 1
LU factorisation, 180

magnetic nanoparticles, 3, 70
mass density, 123
mass matrix, 23, 26
maximal norm, 35
maximum principle, 73
mesh independence, 29
mesh-dependent norm, 140, 149
Method of Manufactured Solutions,
    170

metric, 78, 79
MGVANKA, 59
minimising sequence, 130
minimum energy states, 192
Mountain Pass Theorem, 99
multigrid, 33, 34, 38, 195

natural boundary condition, 136
nematic liquid crystals, 1
nematic-smectic coupling, 123
nematic-smectic transition, 199
Newton iteration, 14, 89, 94
Newton linearisation, 14, 21, 147
Newton's method, 54
Newton-MG, 40, 41
Nitsche's method, 93
non-overlapping triangulation, 43
nonconforming disretisation, 148
null Lagrangian, 10

oily streaks, 179, 189
one-constant approximation, 69
order reconstruction, 74, 76
orientational angles, 106
Oseen–Frank energy, 9, 11, 195
Oseen–Frank energy density, 10, 11
overlap, 48

parameter robust, 33, 34, 50, 61
parameter robust relaxation, 195
PCPATCH, 54
penalty parameter, 149
periodic boundary conditions, 55, 181
perturbed bilinear form, 157
perturbed boundary condition, 185
Pevnyi–Selinger–Sluckin model, 124
phase transition, 11
planar degenerate anchoring, 189
plane defect, 86
Poincaré inequality, 18, 152
point defect, 67, 83
point-block iteration, 45
pointwise boundedness, 25
Poisson's ratio, 91
polydomains, 77
positive definite, 27

positive semi-definite, 180
preconditioner, 19, 35, 46, 53, 198
prolongation, 33, 37, 43, 52
propeller defect, 86
PSS energy, 125

range, 197
real smectic order parameter, 196
reduced model, 76
reference configuration, 91
regularity, 12, 93, 138, 168
relaxation, 33, 43
Rellich–Kondrachov theorem, 130
residual, 35, 37
residual equation, 35, 37
restriction, 37

saddle-point system, 13, 15
Saint Venant–Kirchhoff model, 95
Schur complement, 13, 20, 23, 25–27, 195
Schur complement approximation, 33
second Piola–Kirchhoff stress tensor, 91
second-order PDE, 135
semi definite, 43
Sherman–Morrison–Woodbury formula, 23, 25
shifted deflation operator, 90
smectic liquid crystals, 1, 121
smectic order parameter, 123
smectic-A, 121
smectic-A and -C coupling, 197
smectic-A energy, 122
smectic-C, 121
smectics in a wedge, 199
smoothers, 35
smoothing, 38
Sobolev embedding, 152
solution landscapes, 106
source terms, 170
space decompositions, 44, 46, 48
special functions of bounded variation, 127
spectral inequalities, 46
spectrally equivalent, 27, 29, 35

splitting, 35, 36
splitting norm, 47
stability, 19, 116, 203
star iteration, 44
stationary states, 183
stokes complex, 197
stress-strain relation, 100
successive over relaxation, 36
surface energy, 184, 190
symmetric and indefinite, 28
symmetric positive definite, 11
symmetric traceless tensor, 70, 82, 128

tangential boundary conditions, 84
tensorial nematic order parameter, 196
toroidal focal conic domains, 184, 186
total free energy, 182
traction, 92
triangle inequality, 155, 157, 168
twisted nematic display, 69
two-grid methods, 37

uniaxial, 68, 70, 129
uniqueness and maximum principle, 77
uniqueness of minimisers, 73
unit-length constraint, 3, 13, 20, 30, 195
unit-length constraint improvement, 29
unit-length directors, 9

V-cycle, 37, 38
variational form, 15, 21, 22, 27, 30, 139

W-cycle, 39
weak anchoring weight, 184
weak coercivity, 153, 154, 157
weakly lower semi-continuity, 131
weakly over-penalised symmetric interior penalty, 169
well-posedness, 11, 16, 151

Young's inequality, 204
Young's modulus, 91

www.ingramcontent.com/pod-product-compliance
Lightning Source LLC
Chambersburg PA
CBHW050556190326
41458CB00007B/2067